26.9.

Reservoir Trout Fishing

SECOND EDITION

Adam & Charles Black · London

Bob Church

Reservoir Trout Fishing

Collated and written by
Colin Dyson

A & C Black (Publishers) Ltd
35 Bedford Row, London WC1R 4JH

© Angling News Services Ltd 1983, 1977
ISBN 0–7136–2299–7

First published by Cassell Ltd, 1977 and reprinted 1978
Second edition by A & C Black (Publishers) Ltd 1983

Church, Bob
  Reservoir trout fishing.—2nd ed.
  1. Trout fishing   2. Fly fishing
  I. Title   II. Dyson, Colin
  799.1'7'55   SH687
  ISBN 0–7136–2299–7

Filmset in Monophoto Baskerville by
Latimer Trend & Company Ltd, Plymouth
Printed and bound in Great Britain at
The Camelot Press Ltd, Southampton

Contents

List of illustrations

Photographs in colour
A selection of Bob Church's favourite flies appears between pages 146 and 147

Photographs in black and white

List of illustrations

The text also includes 26 informative line drawings

Acknowledgement

Our thanks to Peter Gathercole, for his excellent work on the colour photographs of the flies and lures in this book, and to John Wilshaw, editor of *Trout Fisherman* magazine, for the loan of some black and white photographs.

To the memory of the late Cyril Inwood, who inspired me more than any other, and to the many fine anglers who, down the centuries, have added to our knowledge and enjoyment of a fascinating sport.

Introduction

One of the most enjoyable jobs I have done, since transforming myself into a full-time angling journalist some twelve years ago, was to write a manuscript for the first Bob Church book, *Reservoir Trout Fishing*. At the time I knew a fair bit about most forms of fishing, and had caught good examples of most species, but trout fishing was (and probably still is) something I had only played at. I regard it as pure relaxation, both from the daily grind and from some of the angling activities I take more seriously. Writing Bob's book, however, helped me to relax rather more successfully! At the time of writing I was, perhaps, in more or less the same stage of development as a fly angler as many who were likely to read the book. I was keen myself to tap the vast knowledge of one of the most effective fly fishermen I had ever seen, and twenty-five years in journalism had trained me to write it all down – hopefully in such a way that others could easily understand. That is not a gift with which too many leading anglers are blessed, and it is no reflection on them. Their knowledge is so extensive that they tend to take too much for granted, assuming that we know much that we do not. Bob was smart enough to realise that, and it was the principal reason why he asked me to help. I jumped at the chance, and while it may be slightly immodest for me to reach the conclusion, the chemistry seems to have worked. When the original publishers had a complete change of policy all their angling publications, and many others, were scrapped. All the angling books except ours were remaindered – the publishing world's polite expression for the knacker's yard. *Reservoir Trout Fishing* escaped for the best of reasons. The original print was sold out, and the few remaining copies of a later, slightly amended version, were being sold through Bob's fishing-tackle business, departing at the same pleasing rate we had seen in the preceding four years. We took it as quite a compliment when A & C Black decided to take us on, and re-publish this considerably revised and updated version. Much has changed since 1977, and a mind as creative as Bob Church's cannot remain at idling speed for that length of time. Methods have been revised and updated, and new waters came into the reckoning. When the book was first written the gates had not even opened at Rutland Water, now, perhaps, the top venue in Britain for the stillwater fly angler. And what neither of us realised in the beginning was that much of what we had written about the bigger waters here was relevant to trout fishing in other countries. That penny dropped when letters started to come from anglers in America and New Zealand, saying how much the book had helped them. One American claimed that Bob's boat-handling techniques were unknown there. That seemed to us unlikely, for angling knowledge travels fast and well. Much of Bob's thinking on lures, for example, was influenced by what he had read of American methods. If he could build on their ideas, it would be naïve to imagine that somebody over the ocean was not deriving benefit from ours. What we did think quite possible, though, was that the majority of American anglers were unaware of the importance of the arts of boat handling. That was certainly the case here in 1977, and even now it is not uncommon to see anglers performing as though the boat is merely a platform from which to cast. It is much more than that. The way a boat is used is as critical for success as the proper use of rod, reel and line. I wondered, when the decision was made to revise and re-publish, whether Bob's approach had changed, for we live too far apart to see each other regularly. I was particularly curious to know whether he had mellowed. He has not. He still has scant regard for the ethics of the chalk stream, which crept on to the stillwater scene at an early stage, and still endure. For him there is no right or wrong way to catch a trout; no unwritten laws to be observed which are within, but different from, the rules pinned up in the fishery office. He respects the rights of others to restrict themselves, if that is what they wish, but for him – and we believe this to be the majority view – the first object is to catch fish. And that means finding out the likeliest legal means of doing just that, on any given day. This uncomplicated approach – albeit backed by vast and complex knowledge of the methods, the quarry and the environment in which it lives – serves him well. He catches hundreds of trout each season, and more than his share of big fish. The one time I have seen him nonplussed was at his beloved Grafham, when the publisher had made special arrangements to take photographs for the front cover of the original book. Sod's law naturally dictated that Grafham would be utterly dead that day. As we flogged away all morning and most of the afternoon the photographer became more and more restless, for he was due on another job. Eventually we had to borrow the one trout we knew had been caught, and fake its recapture. As the misdeed was being perpetrated I hooked one for real. 'This flaming book's transformed you already,' said Bob, before we both collapsed in hysterics at the incongruity of it all. I tell the tale against him to make him sound more human, for that he certainly is, despite a record which looks almost superhuman if you are allowed to read his catch totals. Later that day he bounced right back, when the Sedge popped up on cue. He took a rapid limit of excellent brownies, while I struggled to half his total. 'Back to your typewriter,' he ordered, and here I am again, enjoying myself just as much the second time around. I'm learning again. I'll catch more fish this season than I did last. And between us, I'm sure, we will help others to do the same.

Colin Dyson

The author, weighed down with fish, after his best-ever day at Grafham. The limit catch weighed 31 lbs 7 oz, the heaviest since the very early days of this famous water. The two big brown trout he is holding both topped 5 lbs, and were taken on consecutive casts.

Chapter one

The choice of tackle

Almost anyone fishing the big reservoirs and the better managed small stillwaters stands a chance of hooking a good fish, if he goes often enough. But to catch good fish consistently is quite another matter, for it demands thorough knowledge of a wide range of methods and techniques, and the equipment must be right. To begin, therefore, I will go into the basic tackle requirements first, starting with the rod.

Rods

I am in less difficulty on this subject now than I was in 1977, when this book was first published. At that time the conflict of ideas as to what was right for stillwater trout fishing, which existed among the anglers, was amply reflected by what the rod manufacturers were providing for us to use. There were good and bad glass fibre rods, none of which suited me until I got around to designing my own, with the considerable help of Bruce & Walker. Carbon fibre existed at that time, but it was very early in the development of this material, and nobody had got it right. The best available then were good casting weapons, but they failed, in my opinion, in the important part of their function – hooking and playing fish. They lacked that indefinable quality – 'feel'. I recall hearing that well-known Scottish casting expert Peter Anderson issuing a bristling challenge to an audience. 'I defy you to explain what "feel" is,' he said, and at that time he was trying to convince people that the new glass fibre material was every bit as good as the traditional and cherished fly fishing material – hollow built split cane. Maybe nobody could define it, but it *is* there. We know when a rod is right, and fibre glass did not relegate split cane to museums, and the loving hands of a few traditionalists, until fibre glass blank makers achieved the same 'feel', using different wall thicknesses and tapers. Carbon fibre went through the same process before it became an acceptable, if much more expensive, alternative to glass. Today we have achieved the position where it is possible for me to say that glass fly rods have developed to a state I find it possible to describe as perfection. They won't get any better. Carbon development has pretty well achieved the same condition, and I can leave it to the judgment and the financial means of readers to decide which to buy. There are glass and carbon rods which will achieve everything I shall be describing in this book, and I still use both. The old mistake with glass rods was the quest for the impossible – lightness with power. I was looking for power, and knew I could not have it neatly packaged in something which weighed about 4 oz. I came up with something much heavier for shooting head work which would not only cast the long distances required, but would also bang home a big hook at great distance and, of course, great depth. It was rather controversial at the time, but over the past four or five years anglers have gradually come around to my point of view. The advent of carbon has helped towards that original unattainable goal. You can have a bit more lightness with the power now, but not all that much. The real benefit of carbon lies in the much thinner profile. The slightly lighter weight and the same, or maybe even a little more, power comes in a slimmer package. It is an incalculable benefit in windy conditions, whether one is punching out a line into the teeth of a gale, or forcing the rod back into a similar wind from the rear.

The first carbon rod I tested was a Bruce & Walker job, and its arrival coincided with big winds over Grafham. It was blowing up waves to a fair old height – perfect conditions for perhaps the most energetic and certainly one of the most exciting ways of fishing, Muddler Minnow stripping right in the waves. The method involves long casting downwind, and if it is strong enough to whip up big waves the back-casting problems are easily imagined. I could not believe how well that early prototype was coping. It was slicing through the wind as though it wasn't blowing, yet it weighed just $3\frac{1}{2}$ oz. What's more, it was whipping back a size 9 shooting head, and delivering it downwind to distances I could never have achieved with a glass rod. I was soon to learn, however, that this miracle of lightness had given me a wrong impression of the power. Smashing off on a lure tied to $7\frac{1}{2}$ lb line would have convinced me I had lost a monster had I not seen the fish leap at the moment of impact. It was no more than 2 lbs. I'd struck too hard with a rod far more powerful than the weight had indicated. I had to school myself to strike more judiciously, though since that time improvements with tapers and wall thicknesses on carbon blanks have eliminated the problem completely. I can't see them getting much better, now. In several ways they are better than the best glass rods, and for those scratching their heads and totting up their bank balances the summing up is simple. In flat calm conditions the best carbon rods will slightly out-perform the best in glass, given equal capabilities by the users, but the gap widens the harder the wind blows, no matter what its direction. Depending on the type of rod, the difference in weight ranges from slight to fairly considerable, and that alone makes a difference to the ache in the angler's arm after a long day by the waterside. Take into account the different amount of effort which has to be expended in windy conditions and the argument for carbon becomes even more strong, though I'll repeat what I said earlier – anglers can perform all the functions I shall be describing with glass. They have to decide whether the greater expense of carbon is worth the reward – achieving those functions that much easier and better.

I suppose some readers will be wondering about the third possibility, boron rods, which, at the time of writing, have been around for about two years. I must confess not to have fished with one yet, though I have handled one at a Game Fair without experiencing that familiar 'I must have one' feeling. Expense certainly puts a brake on the old impulse-buying instinct, and lots of anglers who had their fingers burned in the early days of glass stood back and waited a long while until they were convinced that carbon had developed to a satisfactory stage. It would seem they are being even more careful with boron, which has not exactly had rave reviews up to now. I could be proved wrong, but my feeling is that anglers are so happy with carbon that boron won't really take off. Carbon rods have got as light as we need them to be, and we will think twice before paying a lot more for benefits which seem to be quite marginal. For boron to sweep the market the price will have to come down a great deal, and that day seems a long way off. For the moment I'm entirely happy with what I am fishing with, so I'll describe my essential armoury.

There's a 9 ft 6 in rod for general purpose nymph fishing and light lure work. It will take a weight forward size 8 line, and will also throw a size 9 shooting head. The next is the steely 10-footer I was working towards around five years ago, and it's here I've felt more benefit from carbon fibre. There's still no way to achieve the necessary power with this length without also having a bit of weight in glass, though it has come down an ounce or two in the past four years. If I could afford only one carbon rod, though, this would be the first I'd buy. It's for throwing size 10 shooting heads and big lures a long way, and banging big hooks home at distance and, in some cases, depth. For most small water fishing I choose a 9 ft 3 in rod in glass or carbon, designed to throw size 7 or 8 weight forward lines, floaters and sinkers. And for the traditional loch style fishing from boats drifting broadside on I use a carbon rod of 10½ to 12 ft. This style of fishing is really coming back on the big waters, though I suppose there will be some who could claim it has never been away. I learned some of the basics years ago, from the late Cyril Inwood, who was a real master of the method. But then, he was brilliant at everything; one of the truly great fly anglers and the man who influenced my earlier years more than any other. In the early days of big reservoir fishing in Britain, however, there was so much to learn; so many directions we could take. I think the Northampton school of anglers can take credit for a great deal of innovation, and full development of methods now accepted as standard and quite routine, but the old loch style was not a priority. I got more interested when it became possible to compete for places in the England fly fishing team, and what a joy it has been for me. Friends took it up as well, and since people still seem to keep a close eye on what we are doing there are a lot more at it now. Later, I'm devoting a separate chapter to this method, and all I need to say here is that the rod needs to be carbon for lightness and thinness. We work a short line to a bob fly and two droppers, usually size 5, 6 or 7 floaters but occasionally slow sinkers. The only other type of rod I use, occasionally, is a 12 or 13 ft carbon coarse fishing rod which is perfect for dapping. And before I leave the subject I should say that all my fly rods are now fitted with Fuji rings at tip and butt. Those two take most of the strain, but the latest carbon rods have single leg Fujis all the way through. This gives extra lightness and there are fewer flat spots in the curve of the rods. Extra care has to be taken with them though, for single leg rings are easily knocked out.

The right lines

Nothing has developed over the past few years faster than fly lines. I have mentioned the types I use in relation to the rods which throw them, but I must go much more deeply into the subject than that. Carbon rods are quite versatile in the range of lines they will successfully cast, but those using glass must ensure that the weight of line is exactly right for the rod. If the line is too heavy or too light for the rod the angler is stuck with an utterly useless casting unit. That message is as old as fly fishing, but it has not yet been fully absorbed. We still see anglers struggling with ill-balanced tackle, and given the present cost of a day's fly fishing, and the crippling expense of travelling to and fro, it makes little sense.

It's interesting to look back at the original book and note how my own use of lines has changed. Long belly lines have been dropped altogether, and I have no use on stillwaters for the double tapered lines. Weight forward lines, both floaters and sinkers, are vastly better. There's now a neutral density line which hangs just under the surface, which I'll go into a little later, but it strikes me as odd that we still can't buy anything like the home-made sinktip shooting head which I first described in 1977. I would have thought that in the considerable advance manufacturers have made in recent years there would be scope for someone to tackle this job – particularly since sinktip lines are still the deadliest rainbow catchers in my armoury. The sinktip shooting head accounts for a considerable number of these fish, and, since rainbows are favoured over the slower growing brown trout in the stocking policies of most waters, it is essential to have the right tackle. If manufacturers can't give us a full set we'll have to carry on making our own. It is not a terribly difficult job, but my method of doing it has changed a little since I first described it. I now use 6½ yards of heavy belly from an old size 9 or preferably 10 floater, splicing it to 4 yards of no. 7 or 8 tapered slow sinker. I find 10½ yards is about right for me, but there is a little scope for variation. The real plus about the job is that you end up with a free product, providing, of course, that you have the old lines, and most of us do. The only tricky bit is the splice, but if you follow the diagram (Fig. 1) you can't go far wrong. You need to remove an inch of outer cover from each bit of line, either by careful work with a very sharp knife or by dipping the ends in petrol. It is quicker with the knife, but do make sure the central cores are undamaged. Place the cores together and stitch up with a fine needle and nylon thread, and then work in some Araldite, covering all of the stitched section. No need to wait for the Araldite to dry. Using a fly tyer's bobbin carefully whip over the join with fine nylon thread, taking a lot of care to keep it level, and carry on building it up until it is level with the two sections of line. Keep the join taut so that it is all straight when it dries, and then smooth off with a few coats of Polyvinyl. Done properly, this new joint will outlast the line, but do make sure the proportions and line sizes are correct to begin with. The floating part has to be buoyant enough not to be pulled under by the sinking section.

There are four other shooting heads to complete my set – a floater which is mostly used for just one method, fast lure fishing for rainbows in the waves, an exciting and often deadly approach, and slow, medium and fast sinkers to cover every level at which fish can be found. You will be able to pick up some recommended brand names later, along with information about the speeds at which the various lines sink. No need to elaborate here about the weight forward floaters mentioned in the passage about rods, but mention must be made of the once controversial lead line. I think my friends and I were the first to use this American import, and for quite some time we had staggering success. Our catches were so conspicuously better than most other people's that we were practically accused of being on the fiddle. That, plus the fact that we were worried that the lead line might be banned, unless it came into more general use, persuaded us to reveal the big secret. Some of the tweedy types, who regarded any line which sank as some sort of devil's invention, were quite horrified. Others, however, could appreciate that there were certain problems in catching big trout from the bottom in up to 70 ft of water, were they to restrict themselves to dry fly chalk stream tactics. They readily accepted the lead line for what it is – just another superbly efficient way of

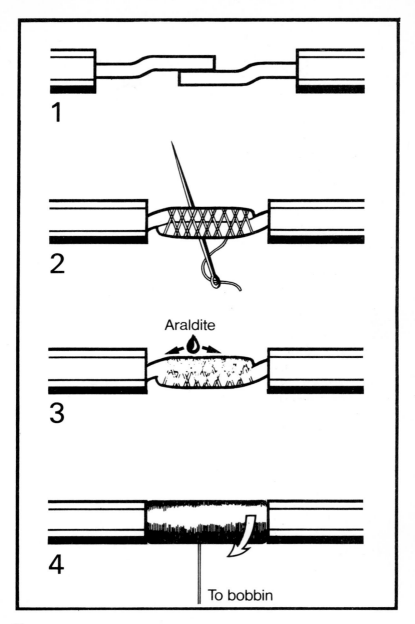

Figure 1
The four stages in splicing.

catching fish which might die of old age without ever seeing a Kite's Imperial or a Lunn's Particular. To judge from what I see on the big waters these days, it is my philosophy which has stood the test of time. Lead cored lines are no longer controversial, except in a few dusty corners of gentlemen's clubs. The reasons why this line works so well are dealt with later, but they can be used as 10 yard, ultra fast-sinking shooting heads, or in full 100 yard lengths for trolling, where this method is permitted. Either way the line has to be equipped with backing and leader, and we do it in precisely the same way as all other lines and shooting heads – the needle knot (Fig. 2). In almost every case the initial hole is made by pushing a pin or needle up the centre of the core and out through the plastic coating. The pin is heated and then withdrawn, leaving a small but open hole through which leader or backing can easily be inserted, and the knot begun. In the case of lead-cored lines the job is easier. You simply take a firm hold of the line half an inch from the end, and with the free hand bend the end around in all directions. The movement snaps the lead inside, and the broken part can easily be worked out leaving, in effect, half an inch of hollow tube. Stick a pin up and make a hole in the outer coating, just where the core starts, again using heat on the pin, and you are ready to start the knot as before. It may not seem to be a particularly secure way of hanging the tackle together, but I can assure you that it is. The outer coating of all

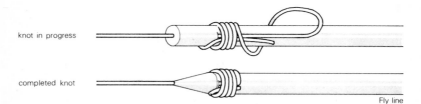

knot in progress

completed knot

Fly line

Figure 2

The needle point in the process of tying (*above*) and complete (*below*). In the lower drawing the pointed finish is achieved by trimming the plastic core with a razor blade and then smearing it with Araldite. The Araldite should be smeared lightly over the complete knot as well, to ensure smooth passage of the knot through the rod rings. This is not shown on the diagram in the interests of clarity.

lines has a much greater breaking strain than the weaker section of leader you are likely to use, and in the case of the lead line the outer coating has a breaking strain of 45 lbs. The finished knot will stand many hours of casting strain, though it goes without saying that there should be periodic inspection for signs of wear. If you see any, cut off and spend a few minutes tying another. It makes no sense at all to lose fish through faulty knots, and the ultimate disaster can be to lose a valuable line as well, though I have never done so through failure of the needle knot.

This, I suppose, is the appropriate moment to consider the type of backing to use for shooting heads, and having tried everything available I'm still convinced that flattened monofilament of around 30 lb breaking strain is by far the best. The best brand now is the pre-stretched Black Streak. It is limp enough not to run into dreadful tangles, and it casts well. While the flattened oval shape is dead wrong for starting the needle knot, because it won't go through the hole, you can easily solve that by sellotaping the end to a flat board and slicing a long point on the end with a razor blade or sharp knife. I can hear some readers asking what is wrong with braided nylon for the backing, and the short answer is not much. It has certain advantages, but for me one overwhelming disadvantage. It is very hard on rod rings, wearing them out in record time, whereas the flattened line seems hardly to wear them at all. The braided line can be greased to make it float, an asset when using floating and sinktip shooting heads, but the flat line floats well when treated with silicone-impregnated cloth. I can remember my writer, Colin Dyson, making that discovery about twelve years ago, when Don Neish sent him samples of his original flat monofil, Tapeworm. Colin was, and still is, more into pike fishing than trout, and he found flat line, well greased, would float for several days, whereas ordinary monofil needed treating several times a day for the greased line livebaiting method. What's more, he reckons he still has some of the original line on one centrepin reel and it's still in good condition. The relatively gentle business of dropping a livebait in the water, and making it swim upwind, however, takes nothing out of a strong line, but the rigours of fly fishing take more rapid toll. Even so, flat backing lasts a long time, providing a close watch is made of the yard or so at the back of a shooting head, which takes some stick in the casting. The final attributes which make me opt for the flat backing are the colour – far less obvious to the trout when lying on the surface – and the more efficient way the flat line helps the sinking shooting heads to present the fly. As you will read later on, I set great store by rapid angle changes for provoking interested trout into taking lures. I won't go into the details now, but there is no harm in introducing, at this stage, diagrams which makes the point, and which also condemn the braided backing.

Fig. 3 illustrates how an interested but suspicious trout will follow a lure travelling parallel to the bottom. Freely-feeding fish will hit the lure at any stage, but the wary trout often allow instinct to overcome caution at the stage in the recovery of line where the lure suddenly turns upwards. It is remarkable how often this happens, but Fig. 4 demonstrates my belief that braided backing does not present a lure in the same manner as the flat backing does (Fig. 5). No matter how much sinking time you allow, the line always seems to ride up at an angle of approximately $45°$. When fish are close to bottom the lure is presented in the right place for a relatively short length of time, and if an interested but wary fish does follow a lure travelling gently upwards there isn't that sudden change which provokes the take.

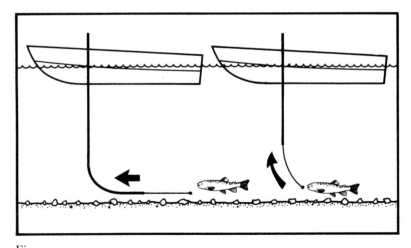

Figure 3
An interested trout will follow a lure running parallel to the bottom, often taking it as it begins to angle upwards.

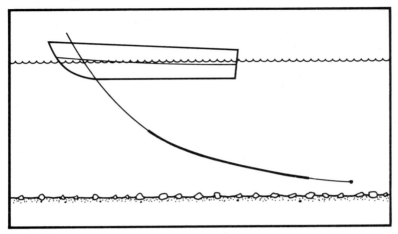

Figure 4
Wrong – the way a shooting head behaves when used on braided
backing.

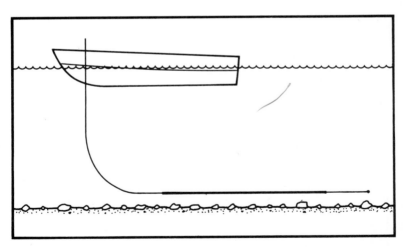

Figure 5
Correct – the recovery angle when the shooting head is backed with
flattened monofil.

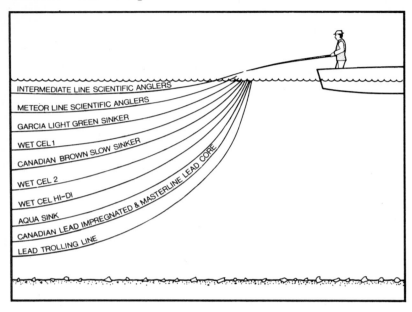

Figure 6
Sinking rates illustrated.

Now it is time to consider the sinking rates of the various lines, nearly all of which I use in shooting head form, often cutting them from full fly lines. Fig. 6 neatly sets out the different capabilities of lines I commonly use, the fastest, of course, shown as the deepest, and so on up. And, while the chart (Fig. 7) is not exactly related to Fig. 6, it does offer a very useful guide to the rate at which commonly used lines sink. The chart was compiled by the aforementioned Don Neish, after experiments carried out in carefully controlled conditions. These are maximum sinking speeds, since at long range the descent is liable to be braked to some extent by the backing, especially if you insist on using the braided stuff. Nevertheless, Don's work gives us some extremely useful guidelines along which to work.

It is, perhaps, a bit of an insult to the intelligence of readers to say that the depth of water often dictates which type of sinker to use – the other key factor, of course, is the depth at which the fish are feeding – but there is some confusion in the minds of some anglers when fishing shallow to medium depths. After all, all sinking lines will go to the bottom. What is required, however, if the fish are feeding on the bottom, is a line which will stop sinking once you start to pull in at the chosen speed. One line will do the job, bringing the lure in close to the bottom without picking up much, if any, rubbish, while another slightly heavier won't, unless the recovery speed is stepped up. If you have to strip in faster than you would really like you are reducing your chances of a take. Many anglers who know this still accept this kind of compromise, perhaps because they simply cannot afford to equip themselves with the full range of lines to cope with every possible circumstance. That is understandable, but there is nothing to stop them building up their collection gradually, starting with a floater, slow and fast sinkers, and then filling in with the rest. There is no route to all-round efficiency, though, until you have the lot.

Before I conclude this section I must make the previously promised mention of the neutral density line. It appears in Fig. 6 as the Intermediate, by Scientific Anglers, and it would be interesting to see some sinking rate figures from Don Neish on this one! I chose one of these, a weight forward 7, for a particular job on a smallish water. I wanted to fish a slow sinking nymph on clear waters where trout were obviously running away from fast sinking nymphs, and maybe spooked as well by floating lines. I

Line Size (AFTM)	5	6	7	8	9	10	
Wet Cel 1	8.0	7.5	6.7	6.3	5.7	5.3	
Wet Cel 2	6.0	5.6	5.3	5.0	4.8	4.5	
Wet Cel Hi-Di	4.2	4.0	3.8	3.5	3.2	3.0	Seconds per foot
Meteor	—	10.4	9.7	9.2	8.7	—	
Garcia Galion	4.7	4.3	4.1	4.0	3.8	—	
Hardy Jet Sink 1	8.0	7.5	6.7	6.3	5.7	5.3	
Hardy Jet Sink 2	6.0	5.6	5.3	5.0	4.8	4.5	
Gladding Aquasink	—	3.0	3.3	3.8	4.1	—	

Figure 7
Line sinking rate chart.

A 12½ lb Avington rainbow which fell to the neutral density line and slow sinking nymph.

found, first time out, that pausing a full minute after the cast, I could still see the line no more than a foot below the surface. Well into the retrieve I suddenly noticed the end of the light green line running off at a right-angle. I'd had no other indication of a take, but the strike firmly attached me to a 4½ lb rainbow. Shortly afterwards, much the same happened with a slightly smaller fish, and then the jackpot for the day – a 12½ lb rainbow which gave a really spectacular account of itself. An Avington stockie it may have been, but I got a lot of pleasure out of it, not least because my combination of line and nymph had fooled three wary fish. They had taken with confidence and carried on swimming. Fig. 8 explains how those takes were registered. Not all big trout takes are shuddering full stops, which jar the arm.

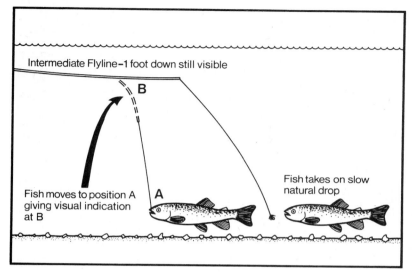

Figure 8
The 'swimming take'. As described in the text, not all takes jar the arm.

Reels

Most writers on fly fishing seem to dismiss the reel as a rather irrelevant item of equipment, the common expression being that it is merely a 'reservoir for the line'. Maybe it is one of the least important items, and it is certainly true that the angler can buy cheap reels instead of the more expensive ones without losing much in the way of efficiency. He still needs to ensure, however, that what he buys is right for the reservoirs. Some reels are rubbish for any kind of fishing, and very few are suitable for shooting head work.

The basic fly reel design has changed very little since they were first used on the chalk stream, and in that era anglers never saw the last third of their fly line after it had been loaded on to the spool. The narrow little reels with small-diameter spindles may be fine for light line work on rivers, but I'm afraid they are bought for reservoir work by anglers who fall for the 'mere reservoir for the line' nonsense. Such a reel will take very little backing beneath the kind of heavy, full line which is usually required for reservoir work. The sad result is that the rear end of the line is

A picture which explains Bob Church's point about the limitations of traditional fly reels for shooting head work. Imagine how the fly reel (*bottom*) would distort shooting head backing into tight loops, and compare it with the wide drum Aerial reel above.

quickly deformed into a series of tight loops which seriously inhibit long casting. Forward taper lines are particularly prone to this trouble, but all full lines for reservoir work suffer similarly on reels which are narrow, with small spindles. For shooting head work the problem is considerably magnified. The nylon backing is distorted into springy loops, and whilst the potential danger can be minimized by stretching before each use, it is a mere matter of time before the owner cops for a real peach of a bird's nest.

Having got that off my chest let us consider what exactly is required. The narrow, small spindle reel is a menace for big water work, and anglers should choose reels for their full lines which are wide, with a fair spindle width. There are suitable reels to be had, usually at the quality end of the market from manufacturers who have tumbled to the fact that casts of the order of 40 yards are both required and achieved these days. I will not name any particular makes because the choice is fairly wide, and you have the option of geared retrieves and automatics as well as the conventional. Final choice depends on personal taste, once the principle of width and spindle diameter is grasped.

From a purely personal point of view I dislike the geared retrieve reels. Small-spindle spools are useless for playing fish off the reel because the recovery rate is too slow, but some of the geared reels fall down for the opposite reason. They are so highly geared that the handle is difficult to turn, even when playing a quite modest fish, and we have to resort to stripping the line by hand. High gearing does give the advantage of being able to reel up loose line quickly when you want to change fishing spots, but in my opinion this small advantage does not justify the extra cost. Other well-known anglers, however, feel differently, and swear by geared reels.

Automatic reels pick line up fast at the press of a lever, but they seem to be declining in popularity. We are beginning to see more of the lightweight reel made by Berkley, a kind of nylon/plastic mix which is very durable, and which is a good product for use with carbon rods. It certainly lightens the casting unit, and there seems to me to be little sense in spending a lot of money on carbon rods and then using a heavy reel with it. Also well worth the extra expense are reels now made from magnesium, which are the right weight for carbon rods. When the first edition of this book was printed there was no specialist reel for shooting head work, but there is now. It's the Line Shooter in pure carbon, a centrepin style reel which takes an 8, 9 or 10 shooting head with about 100 yards of backing. It has a clutch, an optional check and is caged in to stop line blowing off the side of the spool, which can cause kinking problems with the backing. The central spindle is made of carbon, and after using one for three years I have not noted the remotest sign of wear. Neither have I discovered any serious fault in the design. It is perfect for the job, though I'm still quite happy using some of the old style coarse fishing centrepins, which were once greatly superior for shooting head work to anything on the market. Aerial (now back on the market for both coarse and fly anglers to consider), Speedia, Rapidex and Trudex are all suitable because of the wide diameter drums. They don't distort the backing, and for line recovery the angler can switch off the check and bat the line in fast.

The accessories

I will not take up much space on the routine bits and pieces which make up the fly fisher's armoury, for it is mostly common sense. The need for waterproof clothing is obvious, but for fly fishing the waterproofing must be total. You can't hide, like the coarse fisherman, under a brolly. There is no hiding place in the middle of Grafham or Rutland when the rain comes bucketing down. Thigh waders, studded or cleated, are a must. So is a good-quality waterproof bag to keep your gear in. You need a priest to despatch your fish quickly, a spoon to tell you what it has been feeding on and a large rush bag to keep catches in.

The last job before the first cast is to sharpen up the hook – a precaution most fly fishermen neglect completely, at great cost in fish – so buy yourself a good file. You need a tin of silicone grease which, nowadays, I use only for the backing of floating shooting heads. I make dry fly leaders float by spraying floatant on to my fingers and smearing the line with it, for a reason I will explain later on, so get yourself some spray floatant. You also need a small bottle of detergent to make leaders sink. Fly wallets are more convenient than rigid boxes. A wallet will go in your pocket, and the type with a central leaf will hold plenty. Mine carries 300 flies and lures, but I have a large double-sided fly box which is used as the store for my full collection of around 3000 flies and lures.

For bank fishing you want the type of landing net which clips to your belt and shakes open at the touch of a button, but this is *not* the right type for boat work. You need a net on a longish pole – the coarse fishing type is ideal. The reason is simple enough. Trout fight hard at the best of times, but when you get them near the boat they go berserk. The boat is the enemy. I think they see it as the source of all their troubles; the reason for the strange feeling in their jaw and the mysterious force which compels them remorselessly in the direction opposite to which they want to go. Hook-holds wear during a long battle. The neat puncture originally made by the hook can tear into a slit. If a beaten trout is panicked into one last thrash as it is drawn the last few feet towards the boat it can shed the hook. Far better to net it six feet out. . . .

What's next in the accessory list? Polaroid glasses, a pair of sharp scissors, one of those waistcoats with loads of pockets, which are very handy for mobile bank work, and spools of line from 4 lb to 7 lb for making up leaders. I do not use the knotless tapered leaders for still-water work. They are expensive and unnecessary.

For lure work I seldom use anything more elaborate than a straight 9 to 15 ft length of 7 lb line tied to a one yard length of 15 lb line which is needle-knotted to the end of the fly line. If I want to fish a couple of flies on droppers I generally make up a leader from 7 lb, 6 lb and 5 lb line which is joined together via the five-turn water knot (see Fig. 9) which is far and away a better

Figure 9
The five-turn water knot. Two lengths of line are laid side by side. A simple loop is formed and one of the double ends is passed through the loop five times before the knot is pulled tight.

knot than the old and elaborate double blood knot. There is a
dodge to remember about the water knot. When you tie three
pieces of different breaking strains of line together you end up
with a short end of line at the upward (i.e. furthest away from the
point) end of each knot. The dropper flies are obviously tied to
these, and if you keep the droppers short, up to 6 in, they won't
tangle around the leader. If, however, you have been using a
point fly only on a straight leader and you want to add a dropper
you have to tie a water knot with a separate piece of line, and you
end up in this case with two short ends. Always trim off the end
nearest the point and tie the dropper fly to the other end (Fig.
10). It makes sense if you think about it. The downward end is

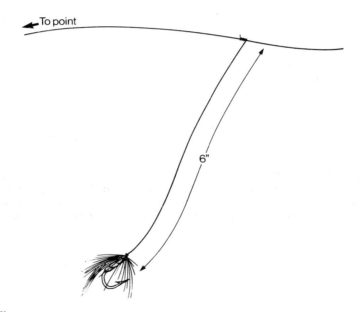

Figure 10
When a water knot is used for droppers the fly must always be tied to the
upper end (i.e. farthest away from the point fly) of the knot. The other
end is trimmed off short. The upper end is less likely to tangle around
the leader during the cast; this is even less likely to happen if the dropper
is kept short – no more than 6 in.

trying to wrap round the leader. The upward end is hanging away from the leader and is far less likely to tangle.

The next and clearly the most important accessory is the fly itself, and I have devoted a special chapter later in the book to the flies which catch nearly all my trout. There are complete tying instructions for each pattern, and in the passages preceding these instructions I describe exactly how each type is used. All I will say at this stage, therefore, is that there are fly patterns which succeed sometimes everywhere, but my list may not include patterns which are particularly successful at your local water. It always pays to spend a bit of time in the fishing lodge, chatting to officials and other anglers, trying to find out which patterns are taking fish. One important detail you may need to know now is the five- or six-turn tucked half blood knot used to tie the fly to the leader (see Fig. 11).

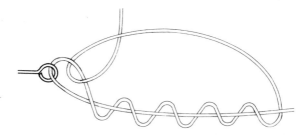

Figure 11
The tucked half blood knot for attaching flies to the leader. The line is taken through the eye and back round itself five times. The end is then passed through the first loop near the eye and then out through the long loop before tightening.

Chapter two

Casting artists – and artisans

I suppose the logical sequence of events has led me to the point where I should consider the question of casting. But if you expect to read the next few pages and finish knowing how to cast, or with an improved technique, then you will probably be disappointed, for I intend to offer very little technical information about casting.

Despite the fact that well-qualified casters have produced some pretty impressive work on the casting art I am left with the firm belief that this is one subject where words fail. However good it is it leaves novices with only a vague idea of what they must achieve, and those who can already cast a line don't benefit either, since they will have acquired bad habits which can be ironed out only by personal attention from a qualified teacher. Casting is a simple art which only begins to sound difficult when one puts it into words and I do not think I would be doing readers any kind of favour by adding to the baffling acreage of words which already exists.

To illustrate my dilemma I will ask a question. If you know how to ride a bike how would you set about writing down, for the benefit of a child, instructions for doing so? Correct. You wouldn't know how to begin. It's the same with me as far as casting is concerned, but most casters learn in precisely the same way as a child learns to ride. Novice casters usually begin with the knowledge that casting has something to do with the weight of the line persuading the rod to act as a spring, and a head full of arm, wrist, head and shoulder movements, not to mention feet positions. Their first attempts to make this 'knowledge' work in their favour are gruesome to behold. For the scraped knees and elbows of bike riding read punctured clothing and/or ear lobes, flagellated nether regions and whip-cracked lines! But it is surprising how soon these tyros learn by their own efforts and begin to cast a reasonable line. From then on they soon progress to the point where they can poke a longish line and catch fish on the reservoirs. There are a few casting artists and a whole lot of artisans on the reservoir scene, and the vast majority are self-taught. It is not an approach I would recommend; I'm just reporting the reality. I would say the quickest and most desirable way to achieve casting efficiency would be to go straight to a qualified instructor. They are pretty widely available these days, either at night school or classes at fisheries, which have a natural vested interest in educating and encouraging their future customers. Read up the subject as well, if you like, but try to store the information in your mind before attempting to apply it. The words will make more sense after a few sessions with an instructor.

If I sat down and scratched my head for a year I could not produce a better casting textbook than *Casting*, an *Angling Times* publication produced by Terry Thomas back in 1960, and people like Dennis Gander and Geoffrey Bucknall have added useful contributions much more recently. But I will repeat my earlier observations that words alone are pretty ineffective. The personal touch teaches volumes more in much less time, though ultimate perfection involves long and persistent practice. To be perfectly honest I have no idea how I might rate as a caster. I am self-taught and my technique has just evolved over the years to the point where I can be accurate, delicate where necessary, and long – right side of 40 yards – in reasonably favourable conditions. Until we sat down to write this book I doubt if I had devoted more than five minutes' thought to the technical aspects of casting in the previous five years, and not much more in the five years before that.

Casting is something which, once learned, is best forgotten if catching trout is your major interest. My own technique is good enough to catch me hundreds of trout. Only if I suspected that I would catch more if I could cast better would I consider trying to polish up a bit. I have no reason to believe that there are deficiencies which amount to a handicap, but there are grounds for at least suspecting that a perfect technique can be a handicap. It sounds crackers, but I will try to justify it logically.

Casting a fly is, in one respect, rather like tying flies. Both occupations can become obsessions; so much so that catching trout becomes a secondary occupation. Some anglers manage both or all three interests perfectly well, and they are pretty well known. But equally well known are the noted fly tiers who aren't much cop when using their own creations and the long casters who catch half as many trout as the consistent 25 yard merchant. These people have devoted too much time to one subject at the expense of the other. As usual there are a few 'hybrids' in the picture too, like long casters who cannot resist lengthening by 10 yards when an audience appears. I have seen these characters catching fish at 30 yards and not catching at 40, simply because their mind has switched from catching to casting. There is a cryptic car sticker about these days bearing the apparently senseless legend, 'Think Trout'. I do not know what it is supposed to mean, but it is not a bad slogan to carry in the head when reservoir fishing. Think about the fly (is it the right one? Is it being fished properly?) and think about the trout (Are they showing? Are they likely to be at this spot if they are not showing? How the hell can I attract a take?). This kind of mental exercise will catch you more fish than thinking about casting.

Once you can cast well there is every reason in the world for forgetting all about it and getting on with some fishing instead. Many top fly anglers will tell you that the moment casting was relegated to a position of secondary importance was also the moment when they started to catch more and better fish. Before I forget about casting for the rest of this book, however, I will mention the types of cast which are required on reservoirs. I will not go into technical detail for the reasons I have just painfully described, but I will remind you of something I said in an earlier chapter – the need for perfectly balanced tackle is paramount.

Nobody can cast if the line does not match the rod. My reservoir casts are confined to three – two and a half if I take into account that I use only the first movement of the roll cast. The basic overhead cast is used when distance is not important, and if it is I employ the double haul. As I stressed earlier, my choice of tackle is mainly dictated by the need for economy of effort and the elimination of excessive false casting.

Most of the time I find myself roll casting the line off or out of the water into the air and belting the line out after only two false

Former world fly accuracy and distance champion Olle Soderblom,
double-hauling in competition. But there's no need to be as good as Olle
to catch good fish on reservoirs.

casts, sometimes employing the double haul and sometimes not,
depending on the distance required. I can, however, get more
economical than that, particularly when using a shooting head
and the double haul. I consider a cast to be just about fished out
when the butt end of the 10 yard shooting head has reached the
left hand. I do not recover any more line, but will then raise my
long rod slowly to a position just behind the head. Takes very
often come during this short movement from fish which have
followed the fly or lure for some distance. I think it is the change
in the direction of the fly, possibly coupled with a change in
speed, which induces these takes. If no take occurs I finish up
with the rod raised and the line bellying from behind to the water

in front – the classic starting position for the roll cast. With practice (and the right tackle) it is possible to poke the rod forward fast enough both to roll the line into the air and send the butt end of the shooting head out a yard beyond the tip ring. That achieved, one good back cast and the correct forward movement is enough to send all the backing out again – a complete cast in just one backward and two forward movements of the rod.

When boat fishing I use the same roll cast into the air, but since distance from a boat is not always so vital I usually follow up with the basic overhead cast (two false casts and out).

Long casting from a boat is required only when the fish show a long way off or when one is fishing deep with a fast sinking line. For obvious reasons short line fishing in deep water will result in incorrect presentation of the fly. Long casting will bring the fly parallel to the bottom for a greater distance and, as you will read later on, the takes usually come during that parallel movement or immediately after the fly changes direction upwards. For this sort of fishing I will attempt to cast maximum distance with the double haul, which is no more than a couple of precisely timed pulls on the line during the backward and forward movement of the basic over-head cast. Doesn't sound much, does it? But when this cast first became fashionable many crack anglers who had long mastered the basic overhead cast took weeks and sometimes months to perfect the double haul despite the coverage in the angling papers. They certainly found the written word to be deficient.

It may help beginners and some more experienced anglers who are not brilliant in the casting department, if I illustrate the rod positions during the vital stages of each type of cast relevant to reservoirs. Fig. 12 shows the basic overhead style, Fig. 13 the double haul and Fig. 14 the roll cast. Space does not permit me to show how the line turns over, but I would recommend anglers to vary, very slightly, from what is shown in the drawings until they become more accomplished. Adopt the more sideways stance shown in Fig. 13 for all three casts, turning the head to watch the line on the back cast, starting the forward movement when you see the line has fully straightened. After a while you will find that you can time it properly without looking, and can adopt the frontal posture shown. In the double haul, which requires two

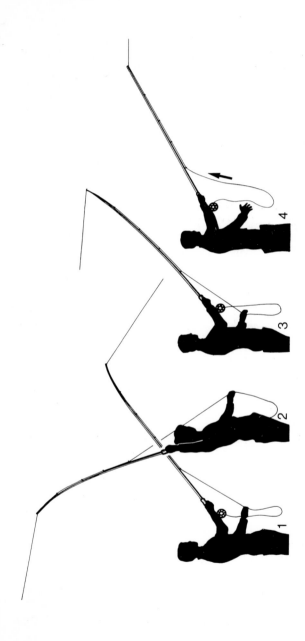

Figure 12

1 With the rod in a raised position, pull in line with the left hand so as to break the cling effect of the water.

2 At the vertical position the rod is stopped to allow the line to straighten out behind.

3 After we feel that little tug on the line as it straightens out behind we now start to push the line forward again.

4 We have line starting to shoot out over the water, with the right hand reaching out and the left hand releasing spare line.

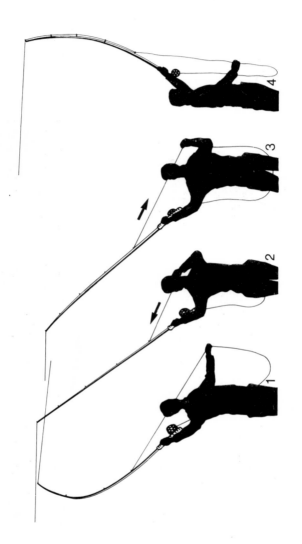

Figure 13
1 The rod is now flexing, and on the first haul the left hand has reached its fullest extent.
2 When the line has straightened out behind, the left hand starts to feed line through the rings.
3 As the forward motion of the cast is started we then start to haul line down again.
4 Here we have the caster applying full power on the forward cast, still hauling line.

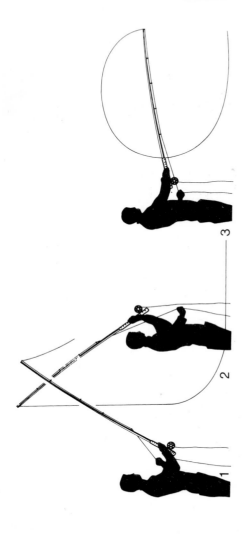

Figure 14

1 To break the surface tension of the water pull in line with the left hand with the rod in a raised position.

2 Take the rod over the shoulder so as to form a belly in the line, which still lays on the water in front. Do this at the same time as your are pulling in line with the left hand.

3 Now snap the rod forward and down thus pushing a loop of line out onto the water.

long and precisely timed forward pulls, the stance has to be more sideways on for the angler to perform those pulls properly.

Casting is all about rod power, line speed, precise timing and that almost indefinable quality, 'feel'. The need to vary a casting rhythm which had been automatic movement for years threw many an angler into temporary disarray. The timing went, and the double haul knocked yards off casts before they mastered it and went on to throw much further than ever before. I would not advise the beginner to attempt the double haul until he can put a line over 20 yards reasonably consistently with the basic overhead cast. How long that might take depends upon how apt is the pupil, how much practice he puts in and whether or not he seeks personal tuition. I see no reason, though, why a beginner could not beat 20 yards well inside a season, and be achieving 30 yards plus in his second season.

Chapter three

Messing about in boats

Terry Thomas once asked me in a television interview what was the real secret of catching a lot of good fish, and without hesitation I replied 'boat handling'. Some unkind gentlemen immediately wrote to *Angling Times* and asked if I was using a trawler! I took it as a compliment, though, for there was a grain of substance in the crack.

For years Northampton anglers were 'docking' with catches which amazed and sometimes upset other anglers. We were so consistently successful that, as in the case of the lead core line, they became suspicious. They knew we were doing something different but, because they did not know exactly what, some of them assumed we were cheating. To put them out of their misery, and to ensure that our methods were not banned by the authorities, we once again spilled the beans. To say we had applied new technology to the business of boat fishing is probably incorrect. It is more like it to say that we borrowed the technology of mariners, ancient and modern, and applied them to trout fishing with enormous success.

In the past few years, boat anglers have become much better equipped, especially at the bigger reservoirs. But it still amazes me that fly fishermen were so slow to catch on, for it was surely common knowledge that there were methods of propelling unpowered craft in directions other than straight downwind! The principle of the rudder is as old as the principle of the boat itself, yet generations of fly fishermen have been prepared to tackle trout with the basic and inadequate facilities provided for them – the boat, a pair of oars, a motor, maybe, and a usually inefficient drogue. Our secret was to equip hired boats with a portable rudder, which we smuggled into the boat in pieces for assembly out of sight of prying eyes, another method of steering across the wind which we called the leeboard, and home-made drogues which easily outperformed those offered at the various waters. You may wonder how this baffled other anglers, as I still do. There we were, tacking and curving about like scruffy Edward Heaths for years on end, while other anglers were restricting themselves to straight lines and one speed, which was often too fast. The advantages we gained were tremendous, and I will explain these before going on to describe the equipment we used and how.

With no other aid but a drogue anglers are restricted to a drift straight downwind, sideways on o⁻ straight down according to whether the drogue is attached at one end or in the middle of one side. But there is another drift method which is more effective.

By using a leeboard to tack across wind you can cast further and recover the fly at whatever speed you want. What's more, the fly is being drawn right across the path of surface or sub-surface feeding fish, usually rainbows, which are moving, as they always do, upwind. The line is coming in with that deadly curve and the characteristic 'whip round' towards the end, which so often induces a take. With each drift being performed under perfect control the scope as far as methods are concerned is considerably widened, too. Sinktip and sunken line fishing can be employed far more often than they can from a boat drifting too fast straight downwind. Another obvious advantage is the amount of time it saves. Imagine how much longer it takes to travel the length of a reservoir in a series of tacking manoeuvres instead of drifting straight down and then having to row or motor back for another drift. The controlled drift just has to produce more fish for that reason alone, quite apart from the presentation advantages. Fishing properly for the maximum possible amount of time is one of the great secrets in all forms of angling, from match fishing through to game fishing.

The same principle applies to our rudder or, as David Fleming-Jones, the recreations manager at Grafham called it, the drift controller. Let's imagine drifting along 100 yards out from a straight shoreline in 15 ft of water. The drift is fairly successful, but we are approaching a big bay. A boat with no rudder has to go straight across the bay. We can touch the tiller and steer into the bay, still working that productive 100 yard contour. The advantages of an efficient drogue scarcely need stating, but the story of how I stumbled upon the design I use now may be interesting.

For quite a few seasons I relied on a cone-shaped drogue made from the frame of a bicycle wheel. The covering was heavy-duty rot-proof nylon. It never occurred to me that it might not be as efficient as possible. We only found out when a pal chucked it into Grafham without fastening it to the boat – the piscatorial equivalent, I suppose, of bailing out at 30,000 feet and discovering that your parachute is still in the plane! The treasured drogue sank to the bottom, and with another boat booked for two days' time I had to knock a makeshift drogue together pretty quickly. No time for fancy shapes. As you will see when you get to the description of how to make it, the temporary drogue was little more than a flat square. I was somewhat apprehensive when I got to Blagdon with it and saw that a strong wind was blowing, but I need not have worried. It was better than the old cone shape in retarding the boat. Other boats that day seemed to be tearing downwind. Mine went so sedately that I could fish a slow sinking line without any trouble at all. I also discovered that the flat design had one major advantage over the cone shape. Try picking a cone-shaped drogue out of the water and it is quite an effort, thanks to the water pressure inside it. With the flat shape a light pluck at one corner collapses it completely.

Naturally my 'temporary' drogue became a permanent item of equipment. It is sensible to make one for waters where efficient drogues are not provided, or where they are permanently fixed to the centre of the boat. That way you are stuck with a loch-style drift, which reminds me of yet another advantage I failed to mention when I was discussing the merits of the leeboard. Rainbows in particular have exceptionally good eyesight, and after surviving reservoir life for a while I am quite certain they associate boats with danger. They can see a boat drifting towards them broadside on far easier than one approaching bow first. Providing them with less to see ensures that more fish come into casting range.

The leeboard

In the days when anglers wondered why our boat was drifting in a different direction from theirs they missed an important clue –

we were taking three seat boards into the boat for two anglers to sit on. I always carry two boards to sit on and one to steer the boat with. They are made from timber 11 in wide and 1 in thick, with 1½ in slats screwed to each end. The stern board is 5 ft 6 in long and the bow board 4 ft 6 in, and the slats are positioned as they are so that the board can be placed across the gunwales without slipping off (Fig. 15). The slats have no function in steering the boat when the board is employed to control the direction of drift. You may wonder why seats are required when they are built into the boat, but the answer, as usual, is simple. A higher seat is much more comfortable and it also facilitates casting.

It was Dick Shrive's idea to use the seat board for steering. He had been reading about the old River Thames sailing barges, commercial cargo boats which have long since ceased to ply the river. He read that when the wind threatened to push them towards the bank the bargees lowered their leeboards, which controlled the drift and stabilized the craft. Dick got to wondering whether such an accessory had any application for fly fishing from boats, and quickly found it had. The wind ceased to become an enemy and became an ally. With offshore winds

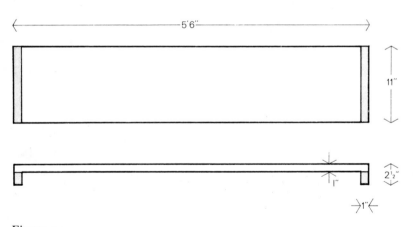

Figure 15
The dimensions of the leeboard used as either a seat or to influence the direction of a boat. Another, 6 in shorter, can be made to fit over the narrower bow of a boat. The shorter version is used only to sit on.

blowing other anglers from the shallow and more productive waters up to 20 ft deep into water vastly deeper and less productive, he had the rich grounds virtually to himself, until he chose to share the secret with us. The leeboard is clamped to the curved part of the bow on the lee side of the boat (the downwind side), hence the name, and you can see from Fig. 16 how it works. The wind is trying to push the boat in one direction, and water pressure on the angled board is trying to force it in another direction. The elements have to compromise; the boat is propelled across and very slightly with the wind. Let's say there is about 200 yards of water where the depth varies from 10 to 20 ft. Beyond that it shelves away to a great depth, where fish can be caught as you will learn later on, but not so often nor so easily. A boat with no leeboard has a 200 yard drift over productive territory in an offshore wind, and it is then carried into no-man's-land. The boat with a leeboard has a more gentle drift of more than 200 yards, if you allow for the angle, and by a simple manoeuvre can be made to tack back across the good ground. You remove the board, spin the boat round half circle and re-clamp the board in the same position on the other side of the bow. *Never* make the serious mistake of trying to turn the boat round with the leeboard still in position, and don't try and clamp it to the upwind side of the boat. It won't work, and it might conceivably overturn the boat.

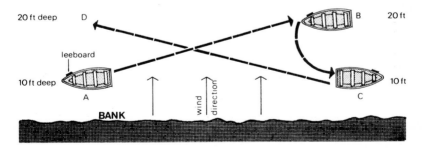

Figure 16
How the leeboard, clamped to the bow of a boat, can make it cut at an angle across the wind. Note the exact position of the board on the bow furthest away from the wind. Always unclamp the board *before* turning the boat for another drift.

Close-up of the leeboard about to be clamped into the correct position. Points to watch: never use the board in winds stronger than light to medium, and never tack across the path of a boat drifting in the traditional style downwind. One major point about the use of boats in general: never, at the end of a drift, motor straight back upwind over the killing ground. It disturbs the fish and infuriates other anglers. Circle back to the starting point in a wide sweep.

The leeboard drift. The leeboard can be seen in position between the author and the bow of the boat. Normally the anglers should face the same way so that their flies will move across the path of their rainbow upwind. Extra care must be taken, however, with the casting, especially by the man in the stern.

The leeboard is not, of course, the answer to everything. It is best used in winds up to medium strength, but it is a very important aid indeed – a vital piece in the jigsaw of detail which, at times, makes all the difference between catching a little or a lot.

The drift controller

The rudder, or drift controller, comes into its own when you want to achieve a measure of control over a longer drift which is more or less in the same direction, as opposed to criss-crossing. As I mentioned earlier it is used to keep the boat roughly parallel to the shoreline a given distance out, or curving round a bay (Fig. 17). It was Dick Shrive, yet again, who tipped me off about the merits, but I still had the problem of making one. I sought the advice of a local garage man, Walt Clark, who had two main qualifications – a practical turn of mind and, more important, a naval background. As an angler, too, he soon grasped what I wanted, and he designed and made one for me. I have drawn the rudder as best I can, so I will now attempt to explain how it is made in some detail.

First the business end – the rudder blade, which is A on Fig. 18. This is $17\frac{1}{2}$ in × $13\frac{3}{4}$ in and made from a sheet of aluminium $\frac{3}{32}$ in thick. B is the main shaft which is made from steel bar $\frac{5}{8}$ in in diameter. The lower end of the bar is slotted to a depth of 1 in. Into that slot is welded a piece of flat metal 1 in × $\frac{1}{4}$ in × 12 in. A further flat piece of metal 11 in long is cut from the same 1 in × $\frac{1}{4}$ in material. The rudder plate is bolted to the welded extension, sandwiched between that and the loose 11 in strip. The bolts are $\frac{1}{4}$ in with wing nuts, marked 1, 2 and 3 on the diagram. Now we move to the bracket which takes the rudder shaft and also provides the means of attaching the rudder to the transom of the boat (No. 4). The bracket is made from $\frac{3}{16}$ in steel plate, 12 in × 6 in with a 2 in lip. The brackets (marked 5) are welded to the plate and drilled to allow the rudder shaft to move freely. The brackets are also drilled out and tapped to take the two winged locking screws. These screws are used to lock the rudder in whatever position is required for a particular drift. No. 6 is simply a safety locking collar. It is screwed up tight all the time the rudder is in operation, and it is there to prevent the shaft slipping and trapping the fingers of the operator when he is setting the drift course. D is, of course, the tiller handle – a 12 in strip of $\frac{1}{4}$ in thick metal, 1 in wide. It is welded to a 2 in × 1 in block of steel which is drilled to fit over the rudder shaft and drilled and tapped to take a $\frac{1}{4}$ in wing screw. The top of the shaft has a flat at the back to enable the screw to lock the handle tight to the shaft. The only further point to make is that the screws at 5

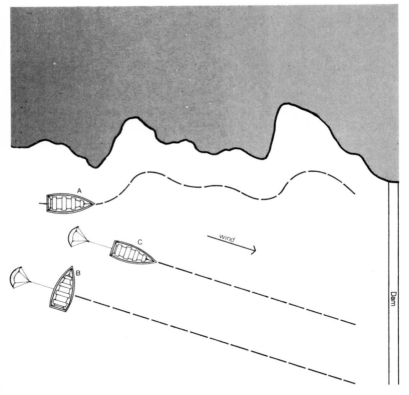

Figure 17

This diagram tells at a glance how the rudder can keep a boat angler in touch with the fish. Rudder-controlled boat A can follow the shoreline and stay roughly over the same depth of water. Drogue-controlled boats B and C are at the mercy of the wind and forced to go straight, perhaps away from the areas where the fish are feeding.

are merely finger-tightened to set the course. When the course needs changing, a touch of the tiller handle does the trick. The screws at 5 remain finger-tight and the rudder will hold the chosen course. It is a simple enough device, but it has stood the test of time. I have used mine for fourteen years without any problems. I am the first to concede that a rather more sophisticated design would be possible, but I doubt if it would work better and that's what counts. The use of wing nuts in my controller enables it to be dismantled for easy transporting, though the original motive was to maintain secrecy by loading it into boats in bits.

Like the leeboard the rudder is best used in light to medium winds. Apart from influencing the direction of the boat the rudder neatly fills the gap between free drifting and the drogue. It comes into its own when, although the wind is strong enough to push an uncontrolled boat a little too fast, a drogue offers a degree of 'overkill' by doing its job too well and slowing the boat too much. A tacking course with the rudder gets round that problem, and it will also help to push a boat a bit quicker on a simple drift in light winds. One word of warning. There is simply no way of using the leeboard, drogue and rudder in concert.

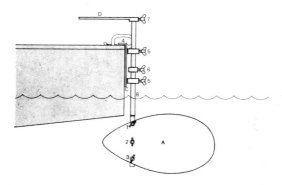

Figure 18
The Bob Church rudder (for dimensions see text). The overall length (not given in text) is a little short of 3 ft. Missing from the diagram in the interests of clarity is the loose 11 in strip of metal mentioned in the instructions. This fits between the wing nuts 1, 2 and 3 and the rudder plate.

Ban warning

Since the leeboard and drift controller were developed to the point of perfection they have been banned at Rutland and Grafham, two of the waters where they were most effective. The Rutland boats, however, do have built-in rudders, a factor which caused a major switch from Grafham by anglers who liked to fish deep. It appeared to coincide with the start of financial problems at Grafham, and I would be surprised if there was no connection. I can think of no logical reason for the rudder ban, which, like the ban on leeboards, followed a certain amount of lobbying from the purist faction. In the case of leeboards the argument was more than just ethical. The safety factor came into it, and I must acknowledge that there is a point here. We used leeboards for

Close-up of Bob Church's rudder, later modified as in Fig. 18. Note the size of the G-clamp. A smaller one would do for clamping the rudder to the boat, but it takes a fairly substantial clamp to hold a leeboard, which is subject to much more pressure. Although the rudders and leeboards have recently been banned at Grafham, the equipment and techniques described in this chapter are valid for most other waters.

The correct position for fishing a rudder-controlled drift – the anglers face in opposite directions and fish across the wind. The rudder clamp can be seen on the stern of the boat.

several seasons without any problems, and so did many others. There were no incidents that I am aware of, but the possibility was there for inexperienced anglers to enter shallow water on a leeboard drift. If the board hits bottom the bow of the boat could be damaged, and possibly there could be a capsize. I remain convinced that the rudder ban was petty, owing more to the fact that anglers who knew how to use them were vastly out-catching those who did not. I am less inclined to quarrel with the decision on leeboards, and I am content to use them where they are allowed. They still amount to a tremendous fish catching aid, and their use must be particularly valid on the big lochs and loughs, and the big waters of Europe, America and New Zealand. It ought to be possible for some inventive soul to come up with a board which is breakable in a collision with the bottom or a pinnacle of rock – similar to the principle of the shearpin on the propeller of outboard motors. Failing that they could be used where the angler really knows his territory, or with an echo sounder as a safety aid.

Meanwhile, we can use other means of influencing the direction in which a boat drifts. In a later chapter on loch style fishing on the drift I mention two of them – leaving the outboard in the water, and somebody heavy sitting as close to the bow as possible, which causes it to dig in and set the boat moving at a tangent. Now if two anglers sit in the bow end the effect is increased considerably. That should set the purists thinking. Watch out for new rules about where we must sit in a boat! I still take leeboards to Grafham, and Rutland, but only as seats. The higher position is more comfortable, and it should also be used when the boat is drifting sideways, as in the loch style.

The drogue

The diagram of my home-made drogue (Fig. 19) tells the story for itself. The dimensions are all there, and it merely remains for me to reveal the materials I now use.

Almost anything tough and rot-proof will do. Nylon and PVC products have the right qualities and I now use rot-proof nylon. Parachute cord, which is thin but very strong, makes up the rest of it, apart from the big snap swivel, which is there mainly to get rid of the problem of line twist. The drogue will turn in the water

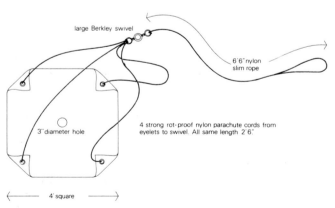

large Berkley swivel

6'6" nylon slim rope

3" diameter hole

4 strong rot-proof nylon parachute cords from eyelets to swivel. All same length 2'6".

4' square

Figure 19
Dimensions of the ideal drogue.

The Bob Church drogue which acts in just the way that can be imagined from this picture – as an underwater parachute.

sometimes, and without the swivel the towline will kink severely. The other use for the swivel is that drogues can be interchangeable. Some anglers make different sizes of drogue to combat varying strengths of wind. If you want to do that use my drawing as the standard type then make one bigger and one smaller.

Whichever size of drogue is used the speed of drift can also be influenced by either lengthening or shortening the towline. Within limits the longer the towline the harder the drogue 'bites'. Whether to use the drogue from the stern or the side puzzles some anglers, but I have no doubt that for the majority of the time the stern is the correct position. Some of the reasons have been discussed previously – namely the business of being able to cast across the wind instead of down it. An offshore wind calls the leeboard into play if you want to fish productive water parallel to the bank. But if the wind is blowing down the shoreline the drogue is obviously required for the boat to go in the same direction but at a speed which permits you to fish properly. If it is

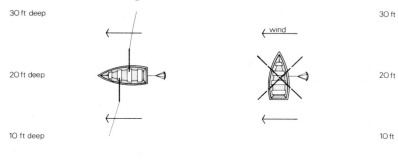

Figure 20
The stern fixing of the drogue allows both anglers to fish across the wind, and cover fish more effectively. While the sideways drift is wrong in the context of the methods described in this section it is correct for the loch style of fishing, which is covered later on, and for surface Muddler fishing.

attached at the stern the wind resistance of the boat is considerably reduced; it acts much less like a sail when it is one end to the wind. Two anglers can also fish across the wind quite easily, either from the same side or from opposite sides, which covers more water (Fig. 20). The side fixing of the drogue dictates that both anglers have to fish a much narrower strip of water by casting downwind. Their range of successful methods is more limited, and they catch far fewer fish.

Fishing at anchor

It is my firm belief that the best sailors catch the most fish but, just like the boys in blue, only for different reasons, you can have a great deal of fun after the anchor is dropped! It is one of my favourite tactics to drift slowly along until a hotspot is located, circle back and anchor over it but, unfortunately, it is rare to find a reservoir which provides an efficient anchor with the boat. A common substitute is a paint tin filled with concrete, but that is utterly useless. The boat will simply drag it along the bottom if the wind has a scrap of strength. If that design was right the Navy would be anchoring their warships with giant bathplugs.

My anchor holds in any wind. It has prongs to bite into the bottom, and linked to that is a long, heavy weight which provides additional holding power. This is tied to about 30 yards of thick nylon cabled rope. Avoid the thin stuff, which is extremely hard on the hands when you are trying to lift a heavy weight. My anchor weighs around 20 lb. The rope is wound round a wooden holder, which is used to act as a marker if the anchor gets snagged on the bottom. I leave it for the bailiff to recover later, in his motor boat. The anchor is usually attached at the bows, especially when the wind is strong. It cuts wind resistance to the minimum possible, and the boats will ride the waves easily without taking in water. Two anglers can fish comfortably with the boat moored in this way, providing they know what they are doing. The trick is for the angler in the stern position to cast across the wind which, of course, must be blowing from his left (Fig. 21). The wind will then carry his fly away from the boat when he is casting, which leaves the other angler safely free to fish straight down with the wind. All this assumes, of course, that both anglers are right-handed. Generally speaking the stronger the wind the

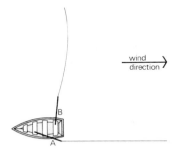

Figure 21
The correct way to fish from a boat anchored from the bow. The angler (A) in the bow seat fishes downwind, and the angler (B) in the stern fishes across wind with A behind him. There is no chance that they can accidentally hook each other though, obviously, they must alternate their casting.

Fishing at anchor. The author, in the stern, fishes the more favourable position across the wind, while the bow angler fishes downwind and behind the man in the stern. Note the position of Bob Church's left hand: he has just completed the second part of the double-haul to give maximum line speed on the forward cast. As the text explains, distance casting is important when fishing at anchor in deep water.

better I like it for fishing at anchor. It seems to put the trout into a feeding mood, and at times some really big catches can be made. You get envious looks in the fishing lodge from anglers who placed too much faith in paint tins filled with concrete!

I suppose that by now I have utterly destroyed the illusion that one of the charms of fly fishing is that the angler can travel light. The bank and river man can, but not the boat angler. When he is properly equipped it takes a while to load the boat, but the argument for going to the trouble is a good one – it takes even longer to unload the boat! The gadgetry described is all essential. I have mentioned nothing which is not vital if you are to achieve the best possible return for your money and gain full enjoyment from your fishing. There are, in fact, certain things I have left out because they are not essential. I have, in any case, exhausted the space available for the preamble, but I can just find room to mention an item of equipment which, though expensive, can boost boat catches enormously – the echo sounder.

A few anglers have used these to locate pike on big waters. As far as I know, however, Yankee Prorok was first to apply this modern electronic aid to locating trout in England. Mike became known as 'the Yank with the television set' at Draycote because

An anchor for an inland sea like Grafham, where waves frequently roll three feet high and sometimes even more, has to be a substantial job. This one is solid iron. Some reservoirs provide paint tins filled with concrete, which struggle to hold a boat in a flat calm!

he used it to locate the ledges around the submerged shoals, and caught trout from them. He reckoned he learned more about Draycote in a day than he could have done in a lifetime of just fishing. It was a completely natural approach for Mike, a former fishing guide in the States, where a sounder is routine equipment. Over here the echo sounder has not caught on, perhaps because we do not have very many vast tracts of water to explore. Even so, it has great potential for the trout fisher, especially if he travels about a lot to strange and different waters.

Chapter four

The hunt is on: 1

I have debated for some time about the best way to describe how to catch trout with the tackle and equipment previously mentioned, and now I have decided that the best way to do it is to take you with me through a typical season, from March to October. By putting events into a logical sequence, and telling both how and when I fished, as well as where and why, a picture may emerge.

If I can get you thinking on the same lines as me, and understanding why a certain fly or method is applied, you will be well on the way to better catches. Naturally I cannot describe every single trip, but I will try to select typical outings and slip in a few outstanding ones as well, covering on the way all the methods which catch the majority of my trout. If you cannot follow me to my waters there is certain to be something there which will help you to catch on your local fishery, be it a major reservoir or small lake. Try to keep things in proportion as the pages flutter by. I shall be mentioning fish which are average to good for the waters I fish, but always remember that there is a standard for each water. A 2 lb trout is outstanding in some places; at others it may have to be 4 lb plus before eyebrows are raised.

Packington will be the first fishery we visit in our conducted tour of the Bob Church season.

Packington (March)

Opening day is very cold, and even though I know the place produced a creditable total of trout last season, and that I'm after new stock fish, I know it won't be too easy. Trout don't feed in the stews when the water temperature is way down, and they don't

One of the lakes at Packington which set the standard for put-and-take fisheries.

change their habits just because they are given a little more swimming space. The destination is Hall Pool. I have fished every Packington water but this Capability Brown-designed lake is my favourite. It is mostly shallow, with deeper water at the dam end, and it is usually very attractive. Not today, though. A cold wind is whipping across the surface and some of that optimism which was bubbling away during the short car journey begins to evaporate. I assemble the lighter of my two rods and equip it with a floating line. There is no need for anything else at Hall Pool, since a floater will cover everywhere from surface to bottom. I pin my faith in a team of three flies which always seem to bring early success at Packington – a Baby Doll on the point, a Claret & Mallard in the middle and a Black & Peacock Spider on the top. Teal & Black, Blae & Black or a Greenwells, all size 12 or 10, incidentally, would do equally well on the top dropper. I cast across the wind because, whenever possible, I like a belly in the line when I'm recovering. It allows a taking fish a vital fraction of slack and assures a good hooking. Fish hooked this way are almost invariably caught in the scissors, and when the hook is there it seldom shakes free. Hooking one today, though, looks like being a real victory.

49

Two hours have gone by and nobody has touched one. Nothing for it but to press on, recovering the flies at the slowest rate possible – the 'figure of eight' recovery method which most people know but which is hard to describe, though I'll have a shot. Forefinger and thumb of the recovery hand grip the line and keep hold as the wrist is cocked slowly back to its full extent. That leaves the line lying across the palm of the hand, and the little finger moves to trap it against the palm. Forefinger and thumb let go and the wrist cocks forward again to take a fresh grip on the line, and the little finger lets go. Then the whole process is repeated, until you wind up with so many loops of line you have to drop them and start again. The recovery has to be slow because trout do not chase about when the water temperature is low, and I stick with the same team of flies because I know that the lack of success is because of the temperature and not wrong choice of fly. I'm hoping to draw the team of flies right across the nose of a trout, so close that it may take a half-hearted snap. The other hope is that the fish will begin to feed when, as always happens no matter what the general conditions, the temperature rises in the afternoon. Both hopes materialise. Out of the blue comes that electrifying signal not felt since last October. The line, which one moment is moving gently towards me, is snaking out the other way. The rod is raised at almost the same instant, and the hook thumps into the first trout of the season. The mind, wandering through lack of action, is instantly alert again. When the sheer elation of the moment subsides comes the warming thought that, given normal luck, the same will happen maybe five hundred or more times before next October. It could be a thousand times, but the thrill of a take and the excitement of the ensuing battle never diminish. It's only a $1\frac{1}{4}$lb brownie I'm gazing at, but it's number one and life feels good. Half a dozen others join it in the bag during an hour-long burst of activity later on.

Other trips to Packington are much the same. The best outing produces sixteen fish, thanks to a buzzer hatch. The first, hooked on a Claret & Mallard, is spooned and it's full of tiny buzzers. A team of green, brown and black buzzers does the rest, the black taking most of them, probably because it closely resembles the naturals hatching. The buzzers are fished in the same way as the wets – allowed to sink well down and recovered slow.

(NB: In March 1976 Peter Dobbs and I took a record catch for this fishery. In a morning's fishing we had twenty-seven trout weighing 44 lb, and all the fish took the Black & Peacock Spider.)

Pitsford (April)

If this book had been written before 1970 I would not be here. There was an enormous head of coarse fish, including perch by the teeming millions. The stocking was pathetic for an 800-acre water, and the perch made sure the small stock trout had little chance to grow. In 1970 the Northants Trout Fishers' Association and the Mid Northants Water Board reached a better understanding; the new Chief Engineer, Alan Simkin, proved very helpful to anglers; the stocking policy improved. Nature took a hand as well. Most of the perch died from the disease which swept the country at that time, and whilst roach and rudd remained they did not seem to harm the trout fishing much. Sport improved dramatically. There are two parts at Pitsford, divided by a road and causeway. The shallow side we call the Puddle, and ninety-nine times out of a hundred we don't fish that. It's the same today. I'm bank fishing the other half, which has plenty of big bays and prominent points.

I choose my favourite Gorse Bush bank because the wind is perfect for a right-handed caster – blowing from left to right. The water is fairly deep close in and long casting isn't essential, but I have the more powerful rod up anyway, equipped with the sinktip line. That choice is made because I checked on the stocking policy, and discovered it had been 100 per cent rainbow. I could catch them anywhere from surface to 15 ft or so down, and the sinktip is good for covering all that depth. Today is overcast and cold so I expect to catch them well down, and the early attack is to be with two lures selected from the range which seems to appeal to Pitsford fish early in the season. The main qualification is the colour – predominantly white or black or combinations of both. Chief ones on my list are the Baby Doll, Appetiser, Missionary, Sweeney Todd, Black Lure and Black Chenille. The first combination, for no particular reason, is a Baby Doll on the point and Black Chenille on the dropper, both on long-shanked No. 8 hooks. I allow them to sink deep before recovering slowly, not with the figure of eight this time, but with the long and steady pull which is more effective with bigger lures. The recovery hand, for the benefit of complete beginners, pulls the line between the thumb and forefinger of the hand holding the rod and the rod itself. When a take is seen or felt you simultaneously take a tight hold of the line and smartly raise the rod. The movement should be smooth and firm; not sharp and jerky. I don't expect too much practice at that manoeuvre today, though. It's a bit on the cold side, and I'll be happy with six takes all day.

The first one comes early, and is missed. There's a lot of casting and walking to be done before the next one comes. I like to keep on the move when bank fishing. The stationary angler frightens fish away by continually covering the same area of water. The rover does not, and he also has more chance of finding fish. I've walked maybe a quarter of a mile before the first success, a $2\frac{1}{2}$ lb rainbow which hits the Baby Doll with a real thump. Next cast at the same spot produces a smaller fish, this time on the Black Chenille, so I know it's not a day for swapping about with flies. Two successes in as many casts with lures which could hardly look more different tells me that location and presentation at the right depth is more important. The two fish came deep, as expected, and so do the three I take later on. Rather better than expected.

Later trips in similar conditions produce similar results, but things are different towards the end of the month when the weather warms up a bit. The first sign is a thumping take on a lure within seconds of the line alighting on the water. There are no signs of the fish surfacing, but they must be feeding higher in the water than they have done previously. I respond to the event by starting the recovery much earlier, and just a bit quicker. The catch starts to climb, and the biggest thrill is to come. In the early afternoon the trout do show on top. They are taking buzzer nymphs just under the surface, and a boil on top gives them away. There are many kinds and sizes of buzzers, but they are mostly black, green and brown. The majority are correctly imitated size-wise by tying on No. 12 and 10 hooks. Off comes the sinktip in favour of a floater carrying a team of nymphs in the colours mentioned. The trick is to watch for a trout to show and then cast into line with it upwind. A figure of eight recovery starts after only a brief pause. The tactic takes four trout in no time, and the pleasing thing is that one of them weighs exactly 3 lb, a good one for the water and obviously a survivor from a batch of 1000 which weighed 2 lb when they went in the year before.

Spooning the fish produces the revelation that it had been fry feeding before deciding to take a buzzer snack. Fry feeding in April! So much for those who vehemently claim that trout don't go on the fry until much later in the season. There are many times when a change in tactical approach is indicated by what is found inside the fish, but this is not one of them. We hear a lot about preoccupation – trout feeding on one item to the exclusion of all else – but I think the problems are exaggerated. Preoccupation *is* a very serious problem for us when trout are gorging a profusion of small insects, like caenis and reed smuts, but when they are on the bigger food items it is nonsense to suggest that they cannot be caught on anything except an imitation. The reality is that trout gorging buzzers can be caught on a lure if it is fished properly. Similarly, trout which are fry feeding can be and often are persuaded to take the offering of a good nymph angler. To hear some people talk trout have I.Q.s as high as university dons, but there is a danger of blinding oneself with science. As always, logic and the balance of probabilities enter the picture, hand in hand, sometimes, with personal preferences. If trout are prepared to take lures or nymphs (as in the case of the Pitsford 3-pounder) the approach is dictated by preference. If there are millions of buzzers about, logic begins to interfere with the preference of a lure enthusiast to the extent that he has to consider whether the buzzer nymph might catch more fish than the lure. I would unhesitatingly choose the buzzer, even though I know the lure would also catch. In the same way it would take wild horses to drag me away from the appropriate lure if the trout are savaging shoals of fry. I rarely allow myself the luxury of irrational bias. I adopt the method which is likely to be most successful – every time.

Ravensthorpe (April)

One of my favourite waters, and one where I always try to fit in at least one boat and bank trip this month. Today it's the boat trip and the method is tried and true. The aquatic life is only just beginning to wake up and stir on the bottom, and the trout will be down deep. They won't be in a mood to chase so they have to be fished for slowly, and the only way to get the presentation right is to fish at anchor. I do more anchor fishing in April than in any other month and, naturally, I try to fish the areas which proved productive in previous seasons. In the case of Ravensthorpe most of them are in a line between the tiny island and the dam wall. The alternative to fishing known hotspots is to fish and move, in much the same way as I do in bank fishing, trying one place for a few minutes and then lifting the anchor a yard from the bottom, allowing the boat to drift about 50 yards downwind before anchoring again. When I fish that way I like to be working a line about 80 to 100 yards out from the bank in 12 to 20 ft of water. When trout are located you naturally stick around and enjoy yourself until things calm down.

Today, though, I'm anchored where I'm fairly confident I'll catch either rainbow or browns or both. This is particularly good brownie water. I rig up with a slow-sinking shooting head and a Baby Doll, which caught most of my early Ravensthorpe fish the previous season. Brian Kench did not realize what he had done when he invented this unlikely-looking lure, but when it became widely used it slaughtered fish on just about every water in the land, and is still doing so. I think it succeeds because, though it is little more than a few turns of fluorescent white nylon wool, the trout can easily see it. They notice it falling through the water and I think they watch it right to the bottom. When it hits bottom and starts to move along, the movement seems too much for the trout, and they grab. This is particularly true at Ravensthorpe, where my April tactic is to let the Doll sink to the bottom, leave it there for a few seconds and then start a slow pull. It is amazing how often the take comes on the very first pull or, sometimes, the second. These are far and away the most telling pulls of the whole cast. Today the tactic is deadly. It accounts for good rainbows and I go back with a limit weighing 20 lb 4 oz – superb fishing. After taking three off the bottom I had switched to a fast sinker. The slow sinker was initially chosen because I had hoped to take rainbows on the drop, but as soon as it became clear that both browns and rainbows were on the bottom the need for the fast sinker became obvious. Why waste good fishing time?

It had not been necessary to employ two other tactics which

work at Ravensthorpe, and elsewhere, for that matter. When the wind is blowing straight at the dam I like to anchor about 60 yards from it and throw a long line downwind. The wind pushes the surface water against the dam, causing it to go down and come the other way along the bottom. Food is carried by that undertow, and the trout are down there waiting for it. Pulling a lure along the bottom always produces good and sometimes spectacular results.

Another method appeals more to some of my friends than to me, though there is no denying that it is effective. They base their early approach on the deep-sunk nymph, choosing to fish from anchor in a sheltered bay where there is a fair ripple, and the water is around 15 ft deep. They fish three nymphs on a leader long enough to fish them close to the bottom. Usually the point nymph is weighted with lead or copper wire under the dressing to help take the leader down. They go so far as to say that this approach is the very best for early season, and whilst I think that is highly debatable I do recognise that it is extremely productive.

I'm in luck with the conditions for my bank fishing trip to this reservoir. The wind is howling straight at the dam and all the other customers have taken one look and departed for calmer climes, but I know better. One of the reasons for my powerful rod is that it will enable me to cast a line through such a wind. It isn't necessary to achieve great distance. The trout are close in taking the food items washed by the current from the cracks in the flat granite slabs which form the dam. A 15 yard cast is enough, and I opt for one lure on a short leader of no more than 6 ft. A longer leader courts trouble when casting into a good blow. This approach, which usually involves fishing with black or white lures, or patterns which mix the two colours, catches so many fish it is almost embarrassing. I go home with another limit and very proud of the best fish, a 3 lb 1 oz rainbow.

Draycote (April)

The mind does some arithmetic as I'm tackling up in Biggin Bay. Nearly 12 000 stock fish have gone in prior to the season, but I'm thinking more about the 6000 rainbows which were introduced last September. About 2500 were caught before the end of the season. Some probably failed to survive the winter, but there must be 3000 or so left, and by now they must weigh 2 to $2\frac{1}{2}$ lb. There is no way, however, to fish selectively for $2\frac{1}{2}$ lb rainbows in water holding plenty of 1-pounders. The only thing I can do is fish in a manner likely to take rainbows and hope for the best.

I rig up with a 3-yard leader carrying a No. 6 Black Chenille on the point and a single black buzzer on a dropper. The wind is right to left, which is opposite to ideal, but it's not very strong. The line is a sinktip which, as I have said before, will take rainbows anywhere from surface to deep down. Today there is no frustrating wait for the first fish. There is a boiling rise only 15 yards out. One cast and half a yard of slow retrieve sees me connected with a 2 lb rainbow and I'm delighted to find it in good condition. Anglers catch a high proportion of black cock rainbows here early in the season. The success persuades me to try a few more casts before attempting to wade. Too many anglers crash in straight away and scare fish they might otherwise catch.

A slow patrol of Biggin Bay provides four more fish before the wind swings round to offshore. The surface within casting range goes flat and the fish move away. Time for a move, and Draycote is perfectly suited to the angler determined to find ideal conditions, having a road all the way round. I drive round to face the wind, remove the dropper and shorten the leader to minimise the risk of tangles from casting into the wind, and soon complete my limit – eight fish making up 13 lb 2 oz.

A subsequent boat trip produces a rainbow limit weighing 13 lb 4 oz but the method is hardly worth describing. All the fish are taken casting across wind on a slow inshore drift, using the same sinktip tackle and the same two flies. Two points are worth amplifying, though. Never begin to retrieve sinktip tackle until the fly has had time to drop from the surface to the bottom, or you will miss out on those lovely rainbow takes which come on the drop. And always re-tie the fly after catching three fish. It may prevent you losing the fourth through a weakened knot. Another thought which should persuade you to be safe, not sorry, is that it is usually the biggest fish which find a weakness in your tackle.

It seems churlish to criticise a water where the needs of anglers are catered for so well, though Draycote's big drawback has nothing whatever to do with the friendly Commander Dunn or his fisheries staff. When the place was built the dams were finished off with lorry loads of big, jagged, tackle-snagging, fly-breaking, bone-cracking rocks. It took four years for this rocky jungle below the waterline to become populated with aquatic life – vast shoals of sticklebacks find refuge and safe breeding places in the cracks and crevices. There are minnows too, and all manner of insect and animal life. The trout cruise the fringes, picking off anything unwary enough to stray out into the open, and it is therefore a rewarding place to fish *if* you approach those fearsome rocks with care.

A method which serves me well, here, is rather unusual – a variation of my cast and walk routine on safer banks. It works only when the wind is running along the dam, for a reason which will soon be obvious. I cast out with a sinktip into water about 15 ft deep and do not retrieve at all. I just allow the wind to belly the line and carry the flies in the downwind direction. I walk slowly along with them, taking great care where I put my feet. The flies are being carried perfectly naturally at the speed of the drift, and a take can come at any time. I usually find I can walk about 10 yards before needing to retrieve and cast again. The method is deadly, but even more so when buzzers are showing in the surface film. When that happens I switch to a floating line with a size 12 winged Greenwells on the point, or a Blae & Black. The leader is siliconed to make it float, and it carries two buzzer nymphs on 6 in droppers which are left ungreased. Trout moving up wind and mopping up buzzers fall for it every time, though the catch rate falls if the buzzer hatch is a really big one. The trout have less chance of finding the artificials among the naturals, but switching to artificials which are bigger than the naturals often works well. Remember that. Whenever there is an abundance of fly life about, offer the trout an exaggerated copy of the natural if an exact imitation does not work. These tactics work well on most waters, so I'll jump ahead to May.

Blagdon (May)

I regard a trip to Victorian Blagdon as a kind of pilgrimage to the birthplace of our sport. The one-time record rainbow stares challengingly down from the wall of the fishery lodge, and a 10 lb brownie in another case reminds you that the four-minute mile of trout fishing is not impossible.

As usual I'm too early for the boat, but I kill an hour and a 2 lb rainbow from the point off Long Bay with a sinktipped Missionary. No fish showing on the surface, so I switch to a slow-sinking line for the boat work but keep the Missionary on. I row for a long, drogue-controlled drift from Rugmore Point back to Long Bay and soon hit a 2¼ lb brown. Two more shake the hook, but a rainbow makes up for it. After that the action ceases for a long spell, and it stays dead until a squally storm blows up in the afternoon. I ride it out at anchor, and find the increased wave action has livened up the fish. A 2 lb brown hits a deep-fished Black Chenille, followed by another rainbow. A 2½ lb brownie on the drift completes an enjoyable but fairly routine day.

The next trip is little different, except for confirmation of something which often happens – black lures take fish when white ones won't, and vice versa. This time it's Black Chenille fished deep which takes rainbows to 3 lb and a 2 lb 2 oz brown, while a companion's Baby Doll is totally ignored. He eventually gets the message and straight away hits a 4 lb 1 oz rainbow, the best of the day. The most productive line of drift proves to be Pegs Point to Rainbow Point.

Chew Valley (May)

The tactics which produced good fish at Blagdon are naturally employed, but half the day goes by without a single take or any indications that a change of method might work, so I plug away with the slow sinker and hope for the best. Companion Jim Collins then notices a modest hatch of buzzers, and switches to floating line and three small flies, anticipating that the trout might start feeding on the top. I'm slow off the mark; not thinking straight today. Jim's rewarded for his perception with a fat hen rainbow weighing 3 lb 2 oz and I'm still on the bottom! A quick switch to the floater brings me four nice rainbows to 3 lb 1 oz and Jim gets another, suffering the terrible misfortune of losing four, all seeming to be around 3 lb. Our fish fell to No. 12 Olive Quills on or just under the surface, apart from one other, a brownie, which chomped a Zulu. All the fish were taken on a drift from Nunnery Point to Moreton Point.

Chew hatches the biggest buzzers I have ever seen, some of them 1 in long and ginger brown. They are hatching during a bank trip on the main Woodford Bank to Villace Bay, and I take five quick fish on the floater, again with the No. 12 Olive. Chew fish are more willing to take off the top than trout in any other reservoir I know, and it's great sport. If 5 per cent of the fish are feeding at or near the top I'm happy to ignore the 95 per cent which are downstairs not feeding, and if fish are feeding upstairs and downstairs I stay up top for the extra excitement of the surface game.

Rising fish at Chew are catchable fish, and there is both thrill and satisfaction in marking a moving trout, casting in front and beyond it and inducing a savage take by pulling the fly right past its nose. It is particularly satisfying at Chew because the fish are big – $2\frac{1}{2}$ to $3\frac{1}{2}$ lb and often bigger still. No other fishery management has mastered quite so well the art of growing stock fish up to 2 lb plus. They go into the waters in excellent condition and they grow very quickly. A staggered stocking policy sees four separate stockings each season, making for more consistent sport. There is no need, now, to join the unseemly riot which usually marks opening day at Chew.

Farmoor 1 (May)

An invitation from friends takes me to Farmoor, the home of some very big brown trout. A bag limit of one brace for visitors persuades me to try for a pair of whoppers, a quest in which chances are enhanced by rules permitting the shaking off of small stock fish and out-of condition rainbows.

Long casting with a slow-sinking shooting head, using lures such as the Badger Matuka, catches plenty to 3 lb. The most effective tactic proves to be to allow the lure to sink right down to the bottom, about 20 ft, and retrieve up the sloping wall. The only lesson from this trip is a reminder that when trout get shy of big lures you start getting tap-tap takes which don't develop. This happens in clear, cold water conditions and the response should be to step down a size. Today I drop from 6 right through to 10.

The trip is most rewarding for a chance encounter with Farmoor's specimen hunter, Syd Brock, whose best four brownies in the last few seasons have weighed 10 lb, 8 lb 6 oz, 7 lb and 6 lb 3 oz. Throw in four over 5 lb and the record is truly impressive. A

The concrete bowl that is Farmoor Reservoir – but don't be fooled by appearances. Farmoor can produce some superb brownies, and the man who catches most of them is Syd Brock, seen in the foreground playing a modest specimen.

man worth talking to, Syd Brock. He tells me his approach is to watch for signs of big fish and to listen to tales of anglers being smashed by something extra special. He marks or finds out the spots and fishes them with relentless dedication. He has the willpower to stick with 2 and 3 in Mylar bodied lures tied to extra long shanked hooks. It is interesting to note that Syd, one of the few among the many thousands of reservoir enthusiasts who have ever tangled with really big trout, chooses to fish with a 9 lb leader. He knows the power of these exceptional fish, and does not intend to lose one after putting so much time and effort into hooking it. It's also interesting to note that he had to make his own rod to enable him to cast shooting heads – mostly fast sinkers – to distances of 40 yards. He reckons Farmoor can offer trout far bigger than he has caught so far, and I'll back him to catch one. I'm too obsessed with the business of catching all trout to restrict myself to his methods, but I don't doubt that his approach would work on any reservoir known to hold big brownies.

Chapter five

The hunt is on: 2

Grafham (June)

Although it seems an age since that first take at Packington in March I have had the feeling that for the last three months I have been merely marking time. Waiting for Grafham to run into top form, as it always does in June, reminds me of how it felt to wait for Christmas in the days when I believed in Santa Claus. The Bristol waterworks lads have no equal when it comes to raising rainbows in the hatchery, but they have no water like Grafham to put them in. Grafham grows 12 in stock rainbows to 1½ lb in just two months. In three months they are 2 lb and by the end of the season those 12 in stockies have become 3 lb bars of silver dynamite. From this month on the lesser waters lose much of their appeal. From here to October, as far as most of this chapter is concerned, I'm staying at Grafham. I'd say every tactic I use to catch at Grafham is applicable at roughly the same time of year to many other waters. The only difference, perhaps, will be the standard of the fish caught. The main thing to remember about Grafham is that it is notorious for lack of surface activity. Fish show less on the top here than they do in any other reservoir I know, and anyone who restricts himself to surface and just under will not get anything like the number of fish that more versatile anglers catch. The other thing to remember is that it is most definitely a lure water. From my experience I will be quite adamant that lures will take the fish two-thirds to three-quarters of the time. That said, we can now go fishing.

This month the main diet of the rainbows is daphnia. There are vast clouds of this minute animal in Grafham, and the rainbows gorge on it. Glutinous balls of the stuff can be spooned from the stomachs of the dead fish and returned, still living, to the water. Curiously, however, when the trout are stuffing themselves with daphnia they are at their most vulnerable to a lure. I think it must send them into a kind of feeding frenzy; willing to snap at anything.

Happily the fishing lodge information before we set out on the first trip is that they are on daphnia. The wind is light and the day is overcast – perfect conditions for long slow drifts in the boat, casting across the breeze with the sinktip line armed with one of three lures which always catch this month and later: Missionary, Appetiser or Black Chenille. The start is typically slow. The boat drifts a mile with not a sign of fish and then wallop – I get three in three casts and so does my boat partner. Then it's all over and the boat is rowed upwind so we can cover the productive part of that drift again. It results in three more rainbows apiece, the best $3\frac{1}{2}$ lb, and every one is taken on a slowish retrieve. The spoon

Spooning a rainbow on the shores of Grafham, against the vast backdrop of the reservoir itself. This fish was full of daphnia – a sign to use a lure, as Bob Church explains in the text.

reveals every fish to be choking with daphnia. The boat has obviously carried us to a spot where the wind and current has concentrated the daphnia. One more drift over it would very likely complete our limits, but who wants to do that by early afternoon? We handicap ourselves by getting out of that area and switching the attack from the obliging rainbow to the more difficult brown. We search deeper water with fast-sinking shooting heads, but the only reward is a throbbing take from a fish which manages to convey the fact that it is heavy before throwing the hook. A switch back to the rainbows also proves unproductive.

Not for the first time, and by no means for the last, we have sacrificed a limit for the pleasure of staying out much longer. Nothing for it but to take comfort from the thought that the score of big brownies would be seriouly reduced if we had always failed to resist the temptation to hammer the rainbows. Slow drifting is the method for the whole month. It will, in fact, take fish right through to the end of the season, but there will be many times when something else will be better.

Grafham (July)

The reservoir has been good in June, but July sees it running into peak form. Limit bags of rainbows are fairly easy to take on the drift, but most of today is spent idly scraping the bottom for brownies without much success. Three rainbows take lures on the drop and one brownie of 2 lb obliges by mid-afternoon. We are in two minds, today, having been told at the lodge that there had been a good sedge hatch the previous evening. Sedge fishing is particularly exciting, and nothing is worse than seeing them hatching when you need only one fish to complete the limit or, worse still, you have eight already.

We anchor in a sheltered bay, pondering the many alternative fly patterns to try. There is really no need to wait for the hatch, since sedges hatch from the caddis grubs which creep about the bottom carrying their little protective houses with them. They are easy prey for the trout, which gobble both the grub and the little bits of stick, stone and grit which form the case. The grub is imitated by the Stick Fly, ideally, but many other patterns resemble it and other bottom-moving creatures. The list includes the Worm Fly, Red Tag, and variations of the Black Lure tied to single hooks. Today we are anchored in water around 12 ft deep and therefore choose to fish with a floating line and a long leader. For deeper water the choice would be a sinktip, but in both cases the method of presentation is the same. The flies are fished deep, with the point fly scraping the bottom, and very slow. We choose to put a Stick Fly on the point and sedge pupae on droppers, one brown and yellow and one brown and green.

An hour's gentle fishing produces two brownies to $2\frac{1}{2}$ lb, but then it really starts to liven up. A few sedges begin to hatch and the odd hump in the water suggests that the trout are now doing what they do nine times out of ten in these circumstances – taking the hatching pupae just under the surface. The tackle needs rearranging, and fast. We have messed about most of the day hoping to enjoy some fishing on the top, but we have left ourselves a bit to do to complete a limit. I need six fish and my mate four, and the rise may well be brief. I choose a well-tried combination of flies for trout humping at the sedge – an Invicta on the top dropper with No. 12 nymphs below, a Green & Brown and an Amber on the point. The leader is greased to within a couple of inches of the point and the droppers are left ungreased. From a boat it is not particularly demanding fishing. Skittering the flies just under the surface on a short line induces three savage takes in no time and the best fish is well worth having – a $4\frac{1}{2}$ lb rainbow. Then we both go through a spell of pricking and losing fish, and in the excitement of the moment we forget what to do about it, which is to increase the size of the flies and retrieve them faster. Conveying the speed in print is impossible. Nobody has found a way of expressing the speed in figures such as yards or feet per minute or second. All I can say is the normal retrieve is quickish and we speed it up a good bit with the bigger flies. It probably induces fewer takes, but the great merit of the method is that fish which are prepared to have a crack at something moving fast really want it. They hit with a bang, and I'd say that 99 per cent of trout hooked like this do not escape. The hook drives deep, for reasons which must be obvious. When we finally do remember and switch to the tactics described, we get five more from six takes. It leaves us one apiece short of a limit, but we go back well satisfied.

Had we fished down deep all day, with the Stick Fly and two

nymphs, we could probably have taken a limit before tea, and the best fish might well have been as big or perhaps even bigger. But we would have missed the extra thrill of catching up top. Pacing the day is important. There are times when we come unstuck, but overall it pays off in terms of enjoyment, and when it does pay there is the additional satisfaction of knowing that the thinking was right.

Before I leave the subject of sedge I will clarify a few more points. The approach we chose for fishing the top was far from being the only one. Other flies would probably have caught fish equally well. The list of likely candidates includes Wickham's Fancy, Silver March Brown, Ginger Quill, the standard Sedge, Silver Invicta, Cinnamon & Gold and a smallish Muddler Minnow. We might also have tried a dry Sedge had the indications led us in that direction. The boiling rise suggested, quite correctly, that the fish were doing most of their feeding under the surface film. Had the rises been more splashy we would have concluded that the trout were taking the hatching fly right off the top, and the correct approach would then have been an each way bet – a dry Sedge on the point with two nymphs hanging below the surface on ungreased droppers. Only if I am quite sure they are taking off the top will I fish a single dry Sedge with no insurance in the shape of the nymphs, but this only happens as a rule in a flat calm. When it does, though, my approach is to cast the dry fly out on an ungreased leader and leave it for a fish to find. I don't retrieve it.

It is as well to remember, too, one of the little tricks which trout sometimes perform. They are capable of having a determined go at the fly and missing it. I'm not sure whether it is bad aiming or deliberate, but I think it is the latter. The aim seems to be to sink the fly with a splashy rise, for they often turn round and hit it under the water a second or so later. The angler who jerks his fly away disappointed by the apparent miss, or half strikes in response to the splash, eliminates the second chance which may present itself. Steel yourself not to strike until line movement indicates a take, and if the splash sinks your fly leave it under for a few seconds before retrieving for another cast.

Grafham (July/August)

Towards the end of July and through August the water temperatures in a normal summer are the highest of the season, and the daphnia blooms are at their most prolific. The rainbows were easy enough to catch in June, when there is daphnia about, but now they go almost berserk. They will take anything that moves, and the faster it moves the more they seem to want it. Caution is thrown to the winds, and the fish can be caught on bright and gaudy lures which would probably have scared the living daylights out of them months before. Why this happens is hard to explain. Perhaps it is because the six square **miles or so** of water are now fully awake. The water simply teems with life, and there are rich pickings for the rainbows, even though they have never really gone hungry since May. They gorge themselves to the peak of fitness, and their extra aggression is perhaps an expression of that fitness.

Rainbows, of course, have a natural aggression towards lesser forms of aquatic life, but if we knew what triggers them to attack we would all catch a lot more trout. There is, however, some evidence that a combination of the conditions and colour does the trick. Earlier in the season it is usually black or white lures which kill. Now it is orange, without a shadow of a doubt, which accounts for the majority of the fish, providing you watch the conditions. I will take you on two July/August trips to explain more fully what I mean by that.

On the first trip a dip of a thermometer gives a reading of 62°F., and a look into the water reveals what you expect to see – algae particles hanging high in the water. Past experience is that the rainbows will be high in the water and afflicted by what we have come to call 'orange madness'. Confirmation is not long in coming. A Whisky Fly, which is as orange as a crate full of Jaffas, retrieved fast just under the surface is walloped by a 3 lb rainbow which advertised its interest by making a bow wave on the interception path! Half a limit each for me and my boat partner is taken in no time, so we pack it in and try for brownies, as usual. There will be no trouble completing a limit later, if necessary, by taking to the Whisky again. Not that it needs to be a Whisky Fly. Literally anything which is predominantly orange will do the trick, flies which fall into that category being the Church Fry,

Spooning a rainbow. Again, the diet in this case is a mass of daphnia.

Orange Muddler, Orange Matuka, Teal & Orange, Dunkeld, Orange Chenille Stickle, Mickey Finn or Orange Wool Doll, which is exactly the same as Baby Doll apart from the colour of the wool. There is no doubt at all that orange fished fast in the top six feet, with either sinktip or slow sinking line, is the colour which triggers rainbows when the water temperature is high. But on our second trip we discover the converse – lower temperatures mean a reversion to June style fishing as far as choice of lure and presentation is concerned.

We find the temperature has dropped rapidly after heavy over-night rain, and the algae has gone. When we try to slip 'em a Mickey Finn it doesn't work at all. Yesterday they ripped Whisky Flies and Mickey Finns to bits. Today nothing . . . until we switch to a No. 8 Black Chenille and fish it deeper and slower. That accounts for three rainbows and two browns which prove to be full of daphnia. They are taken in two drifts over one hot spot, and out of curiosity we try a third drift fishing an orange lure apiece. They are totally ignored. Orange is not the colour which triggers rainbows today. Experiences like this take some of the wind out of the sails of those who claim there is nothing to lure fishing.

Grafham (late August)

Every fish we have caught at Grafham has been spooned to tell us what it had been feeding on, and since May we have been finding odd trout with coarse fish fry in their stomachs. In almost every case the fry has been discovered in fish we have taken deep down, but as August comes to a close fry start to appear in the stomachs of fish caught near the shore. This is a sign that fry feeding has started in earnest and it calls for a change of tactics. Fairly casual deep fishing sessions, mostly tried in an attempt to avoid easy rainbows, have not been particularly productive up to now, but it's time for a serious attempt. We may come unstuck, but we are prepared to risk blanks because the rewards, when they come, are great. The method we employ is likelier than most others to take big fish, and whilst we have caught plenty of trout so far this season the proportion of really big fish leaves something to be desired.

Before I take you fishing for the fry feeders, however, I will describe how our approach is tailored to the situation. Grafham supports teeming millions of perch, bream and roach fry, and the majority we remove from trout are in the 2 in to 7 in class. Clearly we have to fish with lures which are a similar size and colour as the food fish and which have a lifelike quality. We also have to fish where we know the fry congregate. They are not evenly distributed everywhere; they are gathered *en masse* in the relatively few areas which suit them. To solve the last problem first let me tell you that at Grafham there are three areas which are particularly appealing for the fry – the aerator tower, which pumps in water and attracts hordes of fry, the intake valve way out in the middle of the reservoir, which is particularly attractive to fry when there is water coming in, and Savage's Creek, because of the weedbeds and the old hedgerows and ditches. Having settled a few locations for you let me now describe the kind of lures we use.

Normal size lures which have accounted for scores of trout up to now will be perfectly all right for the occasions when the spoon tells us the trout are taking fairly small fry. But when they are gorging themselves on bigger fish a small lure is unlikely to be selected, so we need something bigger – 4 in and 5 in versions of the Appetiser, Muddler Minnows and similar lures. We also find quite deadly a 3 in Baby Doll tied on a long single hook. A key ingredient which contributes so much to the lifelike quality I mentioned earlier is the use of black and white marabou feather which I can take credit for first publicising in this country after corresponding with the American angler, Lee Mitchell. Some commercial suppliers import the feathers of the genuine marabou bird but a useful and almost identical substitute are feathers taken from the inside of the legs of white turkeys, and their coming into use is in my opinion one of the most importan developments in reservoir fly fishing in the past few years. They give greatly added life and movement to a lure, and they have certainly added to my catches since I started to use them. Anyway, let's go fishing and convince you how deadly the fry method can be at its best.

It's very windy, but we don't mind that, since strong summer winds and good fishing seem to go hand in hand. We anchor up in 25 ft of water 500 yards off the north shore and tackle up with fast-sinking shooting heads. The leaders are 5 yards long, for reasons I'll give in a moment, and the fly is a 3 in Appetiser. The cast is a long one, the longer the better, and no attempt is made to recover line until some time after the shooting head has hit the bottom. When it hits the bottom the Appetiser is 5 yards up, and slowly falling head first to the bottom. It can be taken any time during that slow fall, and I have even known rainbows to take when it is being dragged down fast by the weight of the shooting head. However, drop takes are more likely when it is falling slowly, and the first you know about it is when you feel the terrific thump of a hooked fish. More likely still is that the lure attracts the trout's attention during the fall, but it does not take. It watches it to the bottom and grabs at a very early stage in the retrieve – usually on the very first slow pull.

I've wandered a bit; back to the fishing. The shooting head has hit bottom and several seconds have been allowed for the lure to get there, too. No take, so the retrieve begins with a long and steady pull. No take again, so the long slow pull is repeated. That doesn't work on this occasion so the speed of retrieve is increased to fast in the hope that this will induce a lazy trout to wake up and make a chase. We're in luck today, because a $3\frac{1}{2}$ lb brownie hits the lure right at the moment when the retrieve across the bottom has finished and it has changed direction upwards. Almost all the takes when fishing this method come at one of three moments – the slow drop, the first or second retrieve and the lift off bottom. Today we experience all three types of take and more besides, and finish with a double limit weighing nearly 42 lb. The best fish are rainbows of 4 lb 12 oz and 4 lb 4 oz. The gamble has paid off. All the fish are taken from the same spot, in short bursts of three and four at a time followed by long blank periods. We don't move when activity ceases, as we might do with other methods, because this come-and-go feeding pattern is quite typical behaviour in both brown trout and rainbows when they are feeding on fry. They gorge themselves for short spells and retire for a while, presumably to rest up and make room for some more. Moving about, therefore, could be counter-productive. You could try three or four fry-holding areas and be at each one in periods when the trout are 'resting'. Staying put is a better bet providing you are certain you are fishing among fry. One addition to the fry areas already mentioned is the dam wall. It is particularly good for the bank angler.

Grafham (September/October)

Throughout this period most of the methods previously described will catch fish but, starting in August and carrying on through September and into October, yet another method sometimes comes into its own. It is employed whenever there are signs that rainbows are or may be feeding near the top and more traditional methods fail to take them. Daphnia or algae riding high in the water usually means rainbows up top, or you may actually see them moving. But nobody is catching. Our response to this situation is more or less automatic, but not so long ago the method we employ caused a great deal of excitement. It also provoked some criticism from the purists, for reasons I could never understand. Perhaps the great sin of anglers who found the method was to catch fish when others couldn't! I do not know where the credit lies for dreaming up the method. I first heard of it from Pitsford anglers and then saw others doing it and catching at Grafham in a period when my own returns were drooping a little. I merely followed the fashion and caught fish, though it took me a while to get over the surprise.

From a boat the method involves a traditional loch-style drift, the boat side on to the wind and retarded, if necessary, by a drogue. The cast is 25 yards or more downwind with a floating shooting head, and the fly is always a variation of the Muddler. The standard pattern with an unclipped head is deadly. The retrieve is not merely fast, but super fast. The Muddler simply races through the water as fast as you can make it come, skipping through the wave tops like a live thing. It comes so quickly it leaves a noticeable wake, but every now and again a bigger and even more noticeable wake is following on behind, and closing in. A rainbow wants that Muddler, and heart leaps to mouth in anticipation of the take – or the miss. Quite a lot do miss, but it's all part of the fun. When they want it like this the faster you strip in the more offers you attract, though until you have fished this way you probably won't believe it. I think it must be the sheer speed of the lure which triggers them off, for when you try an ordinary retrieve you don't move fish at all on a Muddler. A team of small flies fished slowly may catch the odd one, but not a Muddler or any other lure fished slow. I find the method very exciting. It has brought me rainbows to 4 lb, but never a brownie until dusk. One of my pals has taken a 5 lb brown with the method near dusk. This tends to support the experience of some of the anglers at Ogston Reservoir, in Derbyshire, who found some time ago that certain types of lure stripped fast across the surface very late in the evening produced brownies which were exceptional for the water; far bigger than anything caught in the day. The second hour after sunset was easily the most productive, but the rules were later changed to prevent fishing beyond an hour after sunset. The best catches were taken on warm evenings in July and August, which may suggest that both water temperature and light conditions are an influence where the fast stripping surface method and brownies are concerned.

Grafham (October)

Students of Grafham will know full well that the closing weeks
are very likely to produce the biggest brown trout of the season,
and that it usually falls to a bank angler. The reason is that many
of the big fish have by this time moved close inshore to harry the
shoals of fry, and when they move into range of the bank boys
they are vulnerable. I prefer the greater mobility of boat fishing,
but for the reason just mentioned I tend to run closer inshore than
usual this month. Long, slow drifts not too far out have frequently
located hotspots where trout come for fry, and for our last trip (as
far as this chapter is concerned) we are going to anchor over just
such a spot – under the wood at Savage's Creek. The water is only
10 ft deep and the line is a slow sinker. In most other respects the
approach is much the same as it was for the fry feeding trip
described earlier, and the motive for a virtual return to the
subject is to give you an idea of what the game is like at the very
top of the tree. Consistently catching good fish is satisfaction
enough, but if you fish in the right way in the right places for long
enough eventually you must surely encounter something truly
exceptional. What follows is a true account of an October trip
with Mick Nicholls; a trip which proved a good one for him and
staggeringly good for me. Only Lady Luck knows why I was
selected for additional reward, for we both fished the same way
with the same lures in that 10 ft deep hotspot. But this is what
happened. . . .

It was the familiar short-burst feeding pattern, with rainbows
seemingly the only fish harrying the fry downstairs. In brief spells
through the day Mick clobbered six rainbows to $3\frac{1}{2}$ lb for a catch
weighing over 18 lb, but he did not catch at all until the
afternoon. Within half an hour of anchoring, however, I had
taken rainbows of 4 lb 3 oz, 4 lb 1 oz and 3 lb 7 oz. It was hard on
Mick not to catch in that period, because we had both been using
Appetiser and chenille-bodied Baby Doll lures tied $2\frac{1}{2}$ in long on
No. 6 hooks. He had recommended the size and the pattern on
the strength of his observation that perch fry taken from the
stomachs of trout hooked at that spot on the previous trip had
been $2\frac{1}{2}$ in long and coloured an anaemic white. His observation
and experience paid off for me more than him, but that's the way
it goes sometimes. A week earlier it had been the other way

round, with Mick catching a fine limit of 25 lb, which was much better than mine. A fine angler, Mick Nicholls, and it was no more than justice when he got into the act this particular day during two short bursts of activity in the afternoon. He got three rainbows to my one, and we both lost a few. For me, though, that was the best thing which could have happened, but it was a while before I realized that. Dead hours went by, and we might have moved had we not long ago realized the wisdom of staying put over a fry hotspot. My jackpot fell in the last hour. Mick got stuck into three more rainbows and I got two, going on to complete the limit with brownies in successive casts weighing 5 lb 6 oz and 5 lb 2 oz. I had never performed that feat before, and very few have with fish which gained most of their weight outside a stewpond. The feeling of elation when the catch was weighed in at 31 lb 7 oz – the best from Grafham since the early days – is hard to describe. And adding to the joy was the fact that it won me a free ticket to fish the following season.

Just to bring myself back to earth a bit, though, I must concede that the catch might equally well have been taken by Mick Nicholls, and you may do as well or better some day, if you fish properly over a fry hotspot and the luck goes with you. There is nothing particularly difficult about the method. We fished mostly with slow-sinking lines in exactly the same way as previously described for fast sinkers in deeper water, but we also caught on sinktips and floaters as well.

Rutland Water

I expect you have noticed it. I appear to have gone through the early season on a variety of waters, and through an entire season at Grafham, and not at all to Rutland Water. Blame it on expediency, and on the difficulties of revising and updating a book which are not there when you sit down to write one from start to finish. There was a time when Grafham dominated my fly fishing season from June onwards, and that is amply reflected in this chapter. I have summarised the season at Grafham as it develops, noting how and why the approach changes in response to conditions and events. When first written, the summary was intended to be nothing more than a guide to Grafham, but in checking through it for anything which required revision I realised two things. Everything in it is still highly relevant, and it is more than a mere blueprint for Grafham. It is also a perfectly good guide to fishing on any other big stillwater. I could not have known that at the time, for Grafham was really the only big water we had to go at. Now, of course, we have Rutland, which is vastly bigger, and as the seasons have gone by it has been fascinating to note how history has repeated itself.

Everything which happens at Grafham occurs at Rutland, at more or less exactly the same time. Every Grafham method has its day at Rutland, and if the early part of this chapter is misleading at all it is only in the sense that Grafham still appears to be my favourite water. It is not. It is still the best water I know for growing on stock rainbows, and they alone will always lure me, at the appropriate times of the year, and so will the brownies, particularly at fry-feeding time. But Grafham fell in my estimation, and in that of many others, when rudders and leeboards were banned and when, in 1977, the behind-the-scenes bureaucrats came up with a string of preposterous ideas which would have finished the place for the serious boat angler, and greatly diminished the scope for the bank angler. No need to recall them now, for fierce resistance killed the initiative, if the enterprise can be dignified by that description. The Anglian Water Authority is not noted for expertise in the PR field, and this was perhaps its most disastrous exercise. The memory of what they tried to do has lived on in the mind, and even some anglers who did not feel too strongly about leeboards and rudders voted with their feet. Had there been no suitable alternative I suppose it would all be over and forgotten now, but there was an alternative, thoughtfully provided by the Anglian Water Authority – Rutland Water. Maybe the march from one river division to another was an empty gesture, at the end of the day, influencing only the amounts by which both waters sank into the red. These are tricky times for our sport, with many beginning to think we have gone too far in the provision of stillwater trout fishing. There seems to be an over capacity, a problem inflamed by the undeniable fact that anglers are less able to pay the prices some fisheries feel obliged to charge. It would be a disaster for the sport if the Water Authority cannot afford, at some future date, to continue to offer the present high standard of sport at Grafham and Rutland. And Rutland is good, despite the periodic grumbles from anglers who cannot catch there. It is good enough to supplant Grafham in my affection, anyway, though I am the first to admit that it offers a somewhat harder challenge. At 3000 acres it is vastly bigger than Grafham, but the managers have risen well to the stocking and restocking challenge. However well they do it, though, anglers must realise that the fish vote with their fins, just as anglers can vote with their feet. There are vast areas where the fish choose not to be, so Rutland presents us with a rather bigger than usual location problem.

The basic guidelines are the same, though, as they are for Grafham. Rutland's fish will still be down deep when it is cold, up top when it is warm, and at various levels in between according to time of day and other factors, such as light intensity and the whereabouts of the daphnia. They will still be most vulnerable to the bank angler at dawn and dusk, and most available through the day in areas where the wind is blowing whatever is on their menu at the time. The same underwater features which attract fish in any water are there at Rutland, doing exactly the same thing. Watercraft, experience, knowledge of the quarry and its habits are just as much the key to Rutland, as to Grafham or anywhere else. It helps, however, to know that about 90 per cent of the fish inhabit the vast but relatively shallow (15 to 25 ft) areas at the bottom ends of the North and South arms. I have dredged the memory for anything which has worked for me at Rutland which is peculiar to that water alone, and have come up with nothing.

The one method which has not been covered in detail with regard to Grafham or anywhere else is trolling with the lead line, and since Rutland has a large area of water in which the method is allowed this is, perhaps, the best place to describe it. One quality which draws me to Rutland is the high standard of the brown trout fishing, and trolling with the lead line is an excellent way to catch them. It is not, of course, the only way, and maybe not the most enjoyable way. The big brownies will fall to the methods described in detail for Grafham, but the time and circumstances might not be right to use them. My thoughts turn to the troll when there's nothing happening up top, and my experience of the conditions suggest that the likely action will be down near the bottom. I also troll, sometimes, when we have had a few early fish and it has gone quiet. A few hours devoted to big browns, while waiting for a bit more surface action in the evening, are never misspent. The potential rewards at Rutland are so great that I am tempted to troll, sometimes, even when the rainbows are going well.

First let me clear up any confusion in your minds about the difference between legal trolling, in the prescribed area, and methods which may seem markedly similar, which can be legally employed where trolling is banned. If you are rowing a boat and trailing lead lines, deep down, at a speed determined by the oars, you are trolling. If you are doing the same thing but on a wind-controlled or drogue-controlled drift you are not trolling, and can employ the method where you like. The difference may seem academic, but in fact the methods are quite dissimilar. Wind and drogue drifting rarely achieve the right sort of speed, which is one of the critical factors in the success or failure of trolling, and the drift method is normally most effective when following the 25 to 30 ft contour. Real trolling takes in depths vastly deeper, and one part of Rutland is 110 ft deep. This is in the main body of the reservoir, where trolling is legal, but unfortunately most of the big brownies spend much of the season in the North and South arms. They come into the trolling area in greater numbers towards the end of the season, but there are usually enough fish around to make trolling a viable proposition most of the time. The method seems simple enough, but like many other things in fly fishing, appearances can be deceptive. The challenge is to get a big lure travelling just off the bottom, at a speed which the brownies find acceptable. There is no precise formula which tells us how to do that, and we have to rely quite heavily on trial and error, plus mental messages about how we caught in similar conditions in the past.

The line we are using is the full 100 metres of American lead core, mentioned earlier, which changes colour every 10 metres. That colour coding is an invaluable asset, for we can tell exactly how much line we are letting out. But exactly how much do we need to let out? That's the tricky bit, for it can vary from around 40 metres, depending on depth and speed. Obviously an echo sounder is a major asset here. In some ways we have to wait for the fish to tell us whether we are doing it right; if we aren't catching we are very likely doing it wrong. If nothing happens after a reasonable length of time we can try altering the amount of line out, and rowing a bit faster or slower, until a nice fat brownie tells us we are finally on the right lines. The trollers on the big Lake District waters, who have fished for char in the same manner for centuries, have a totally different approach, trailing spinners off a line held vertically by a large, axe-shaped lead, and their rods are held out at an angle to the boat in special clamps. Their problem of finding the feeding depth is solved by the multiplicity of spinners, each fishing a different level. None of that for us, and while I suppose we could borrow their ideas on propping the rod I'm not sure it is the right way. If you are rowing a boat and trailing up to 60 yards of line you have the problem of striking. Ideally you want the fish to hook itself, and positioning the rod so that it can bend when a trout hits the lure is not likely to assist that aim. I place my rod down the length of the boat, with the reel – clutch screwed up tight – about an inch clear of the seat and the rod point beyond the stern. Hopefully the nearest rod ring at the angler's side of the stern will be further away from the woodwork than the reel is from the seat, for, when a trout does take, the first the angler knows about it is the hard thump of the reel being jerked forward to hit the seat. If the ring hit first, it could be damaged. Having everything set so that the trout hits very solid and instant resistance results, on most occasions, in a hooked fish.

I suppose you will have gathered by now that while there is a degree of sophistication in relating speed of boat to the depth, and the amount of line out, the trolling method is basically crude. Deadly fish catcher it can be at times, but fly fishing it ain't! Even the flies we use are exaggerations of what are normally used in more standard forms of lure fishing. That White Marabou Muddler which works at fry-feeding times (as described in the Grafham section) is a killer with the trolling line. So are the Appetiser and Black Chenille, tied in tandem, along with a big yellow job we have christened Goldie. Mention of that one calls to mind a slight complication in the trolling technique, which I uncovered during an outing at Rutland with fisheries officer David Moore. We weren't after brownies. David was wondering where his larger rainbows had got to (rainbows have the unfortunate habit of disappearing at Rutland, never to be seen again) and he asked me out to try and catch some fish he had located with his echo sounder. The little specks on the screen

A typical catch of Rutland brownies, all around $3\frac{1}{2}$ lbs.

suggested a concentration of fish 30 ft down in 80 ft of water, and David was hoping they would be his absent rainbows. Out went the lead trolling lines, armed with Goldie lures, with enough line out to reach 30 ft, we hoped, and bang – a string of brownies, all around 3 lb to 3 lb 6 oz. Until they saw our lures they were just hanging there, on the fin, though stomach contents of mainly hog louse and shrimp told us they had been in the margins earlier in the day. The suspicion that they would very likely make a return trip to do the same in the evening proved correct, for I ambushed more of them later on, again all in the 3 lb bracket. I'm not sure what practical lessons there are in the discovery that the brownies would almost literally sit around above mid-water in 80 ft. Maybe they always do that between feeding spells, and it could be why working the 25 to 30 ft contour is often so effective through the day. It may be that around 30 ft is their ideal resting depth, and fishing at 30 ft over all the deeper sections may prove to be the next of the 'in' methods.

The use of the echo sounder as a fish finder as well as a depth indicator may not find favour with some people, but my general creed of making use of anything which is legal leads me to ask why not? It's something the Americans have done for years, but they are a practical race, not hidebound by traditions which existed here when they were fighting the Indians. Their pragmatic approach is more relevant to our stillwater fishing than that of our trout stream purists.

One final word about Rutland. As previously mentioned, the boats have built-in rudders, enabling us to perform most of the functions they stopped or hindered with the leeboard and rudder ban at Grafham, but the angler needs some means of jamming them in the correct position. He also has to be able to alter the setting easily and quickly, and the best way we have found is to stuff thick foam rubber between the rudder arm and the stern board. It holds the rudder firmly for any given direction, but it is flexible enough for any course correction to be quickly achieved.

Chapter six

Variations on a theme

The need to maintain some sort of logical sequence on my trip through a typical season caused me to leave out a great deal, as experienced readers will no doubt have recognised. My motive was not to complicate things too much. Cramming every single method into that section, and hopping about from one subject to another, would perhaps have disguised the fact that there is a basic approach to the game to which one adheres most of the time. A typical season, however, is never quite that simple. I often find myself departing from the normal either in desperation, on difficult days, or in response to events which are not quite normal. I suppose the lesson I will be trying to preach in this chapter is the need to be versatile, but there is no way I can put the message into logical sequence. The very nature of the subject prevents that, so I will have to tell the story by describing a series of disconnected incidents which have occurred at a wide variety of waters during the past few seasons. . . .

Draycote (second season)

Catches taken on conventional methods are dropping. Anglers are complaining about the poor sport; my friends and I are saving our breath and wondering what to do about it. We know the head of brown trout is huge; we know they are not being caught; we wonder if they have retreated into depths so great that anglers are unable to get at them with the means at their disposal. We make shooting heads from lead-cored American trolling line and venture into the unknown.

The very first drift in extremely deep water produces six brownies between $2\frac{1}{4}$ lb and $3\frac{1}{4}$ lb on No. 6 Black Chenille. Further drifts run the total to a double limit of browns to 4 lb.

Two more trips produce the same results. In the same period nobody is catching, so we spill the beans and the Draycote grumblers are silenced. The method takes trout from depths up to 70 ft close to the valve tower. Manufactured fast sinking shooting heads also catch fish but not as many. They take longer to sink and have a tendency to lift off bottom on the retrieve. The 'lead' gets down fast and hugs the bottom on the retrieve. The method therefore becomes standard for fishing all deep waters known to hold a good stock of brownies, and is called into use when conventional methods fail or when rainbow fishing is too easy and we want to top up the catch with a couple of good fish.

Draycote (third season)

Simple arithmetic and other factors persuade us that by now this water must hold some really big brownies, so we devote a lot of time to the lead line method developed the previous year. It continues to take good browns until, one day, I feel a terrific thump in 30 ft of water near Barn Bay. After the take nothing is felt for a few seconds. Then the water simply explodes as a big brown rockets up like a Polaris missile and hangs, four feet in the air, in a shower of spray. It is still connected to a Black Chenille, but only briefly. The fly shakes free as the fish crashes back onto the surface, but it's not the end. The fish has damaged itself with that rapid run from great depth. It flounders near the boat, goes under and surfaces again, a little further out. It goes round the boat in ever increasing circles and finally disappears.

Some time later that day a bailiff sees a fish floundering in difficulties near the Barn Bay shore. He fishes it out and weighs it . . . 6 lb 9 oz. No bigger fish is taken at Draycote that season. I lose not only a big trout but a free season-ticket prize and the satisfaction of achieving the ultimate reward for pioneering a method. . . .

Tittesworth

Closest English approximation to an Irish lough is Tittesworth, in my opinion, and its fish seem to have the same idea. Nowhere else have I found brownies which fight harder than they do in this crystal clear fishery, which nestles majestically in the lap of the

Pennine hills near Leek, in Staffordshire. On my first trip I make a bit of a fool of myself, announcing as at least 4 lb a fish which turns out to be just 2½ lb. All I can say in my defence is that it felt that big.

Fishing close to the rocky banks, some of which are accessible only to the boat men, seems to pay at Tittesworth. Regulars say the Missionary is one of the top lures, and I saw it account for a 3½ lb rainbow which fought every bit as hard as my brownie. There's a big buzzer hatch, and from June onwards the fish rise avidly to dry flies.

Damerham

After a good start we are in trouble because of an event we have not encountered on an English stillwater since the old days at Eyebrook – a mayfly hatch! We pay the penalty for not having anything in our wallets looking remotely like a mayfly, and we catch nothing as the trout splash happily at a rare feast. Eventually I notice they are now taking the spent flies, which are lying motionless on the water, wings outstretched like an aeroplane. A flash of inspiration tells me to try a Church Fry with a squirrel-tail wing. I part the wing down the centre, and grease each half with silicone, twisting the wings like old waxed moustaches and forcing them apart. The ploy works immediately; a limit catch is quickly completed. The offering looked nothing like a mayfly, spent or otherwise, except in the shape, but on this occasion the right silhouette was more important than the colour.

Queen Mother Reservoir

This Datchet fishery is unique among the 'concrete bowl' waters. It's so much of a bowl shape that anglers can't stand on the banks, and the fishing is boat only, and pretty simple. Main choices are to anchor the boat 40 yards out from whichever part of the bowl the wind is blowing at, fishing sinking lines and offering lures close to bottom, or fishing on a drogue controlled drift. If the latter method is chosen, it still pays to keep around 40 yards, for the central depths seem relatively barren. I feel the drift method is best here until fish are located, and the depth dictates

Bob Church with a 5 lb brownie from Datchet.

a fast-sinking shooting head. On various trips I have done well with the Appetiser. They certainly seem to sort out the better rainbows, perhaps because sticklebacks are a major part of their diet. The large brown trout here have all grown on from 10 in stock fish.

Eyebrook

Nobody is getting so much as an offer, let alone a fish. Nobody except Cyril Inwood, who seems to be in his mischievous mood. The Inwood total mounts remorselessly; still nobody catches. 'What are you using, Cyril ...?' The answer floats across the wind: 'A claret-coloured nymph. ...' Everybody switches to a nymph as near claret as possible; nobody catches. The Inwood total amounts to forty; nobody else catches. We grab him as he wades from the water and discover his nymph isn't claret but pillar-box red! 'Must be colour blind,' the old fox mutters with a wide grin on his face.

The tale becomes part of the Inwood legend, and pillar-box nymphs, which presumably imitate bloodworms, become part of the standard approach for Eyebrook. Nobody ever again beats the field forty–nil, but then, there was only one Cyril Inwood.

Lower Moor Farm

A 30-acre gravel pit at Oaksey, near Cirencester, which offers very good fishing. On one trip I have a very good morning with fairly routine methods, and during a quiet afternoon examine a shallow bay where, I had been told, brownies had shown the previous evening. I could see why. The shallows were simply heaving with corixa, and I sit there at the appropriate time, a Corixa imitation on a floating line, waiting for a trout to show. Eventually one does, about 20 yards out. I'm too slow to react, but I make a guess at the direction and ready myself to cast at the next surface sign. The Corixa drops right in the ring, and I'm into a 3 lb brownie. Inside ten minutes I've done it again. It's this sort of variation which really makes a day.

Pitsford

The 'horaria' is up, and not for the first time I wonder whether 'horaria' is the origin of the word horror. 'Horaria' is one of the six species in the caenidae family, and everyone knows how difficult trout are when they are obsessed with caenis. We chop and change about with small wet and dry flies, and concentrate on the dry when it becomes clear the fish are taking right off the top. We scale down to a No. 7 long belly line and a fine 3 lb leader, but no joy until I realise the trout must be seeing the silicone greased leader. We swap for ungreased 3 lb line and lightly touch it up with fly floatant sprayed on to finger and thumb and then smeared on the line. Treated thus it floats but is hardly noticeable, and we catch fish on size 16 Tup's Indispensable and Grey Duster. A gently inched retrieve gets the best response.

Next trip the trout are taking caenis in a flat calm with bright sun and nothing works, not even a fast-stripped Muddler, which can be the answer on some occasions. Later a stiff wind gets up and we try loch-style drift tactics without result. Sheer desperation drives us to offer them a different fly action by switching to a leeboard-controlled drift. Casting 30 yards across the wind with a sinktip and a Black Chenille lure we do not recover any line. The boat, now drifting across wind, pulls the lure round in a half circle. Just as the fly swings quickly round the last bit of the half circle before straightening we catch a rainbow which is stuffed with caenis. The tactic produces more fish, and may again the next time they are taking caenis on a windy day. Then again it might not, but we can try it before admitting defeat, as anglers usually do, and go down deep for brownies instead. That move, however, will probably remain the standard answer to the caenis problem!

Church Hill Farm

With a name like this, how could it go wrong as a fishery? It hasn't, ever since it opened, and is one of the most innovative of the smaller waters. Hot lunch available from 1 p.m., for instance, and it's good enough to denude the banks of anglers at around that time. Very civilised, and maybe a subtle conservation

A brace of Church Hill Farm rainbows weighing 18 lbs 2 oz.

measure too! Not that this Buckinghamshire fishery is averse to anglers catching. They stock with rainbow, brown, brook and tiger trout, with some rainbows in double figures. I've seen three come out in one day, and one angler with a brace weighing 18 lb 2 oz! No particularly memorable angling message has emerged for me from this fishery – nothing which has not been common to many other waters – but it is a shining example as a quality, small water fishery. They add a lot of variety to our seasons.

Grafham

It's in the difficult midday period and we are performing Cyril Inwood's 'show 'em and move it' method of short-lining with dry fly on a loch-style drift (see next chapter). Our motive for doing it is hardly a good one – we simply don't know what else to do! Several methods have failed, and we started the loch style because odd fish were showing on top in light breeze conditions. The only trouble is they are definitely not showing near the boat, and it suddenly dawns that they are perhaps able to see the boat more easily now than they can when there is plenty of wave action.

We switch to drifting point end first, with the aid of the rudder, at the gentle pace generated by the breeze. For some reason we also opt for the sinktip line and short, lazy casts of not more than 15 yards across the wind. I suddenly start catching on two buzzer nymphs, and a mate does the same on a lure/nymph combination. The trout are taking 5 to 8 ft down over all sorts of depths of water. They are feeding on midge pupae but are prepared also, apparently, to take anything providing it is presented properly. Short-lining with the sinktip was born; as easy a method as you could think of, but it has since become one of our deadliest. It involves no more than a casual cast, and the allowing of a small belly to develop in the line before following the line with the rod on its semi-circular drift. After that a gentle retrieve and a careful lift-off and re-cast. A feature of the method is the confident take it provokes from rainbows, which rarely come off when hooked. This is in sharp contrast to the situation when loch-style drifting. A high proportion of trout come unstuck that way. I think one reason why short-lining works well in light winds is because the rudder helps push the boat through a little bit quicker but, of course, if the wind strengthens the drogue should replace the rudder. Short-lining with the sinktip is similar to the longer casting across wind with a sinktip which proved one answer to the caenis problem at Pitsford.

There is another approach with a slow sinking line which works in very much the same way from a moving boat. A long cast across wind is followed by a pause to allow the line to sink a little. The boat's drift causes, at the same time, a bow to develop in the sunken line. At the start of a slow retrieve the lure travels for a short distance in the opposite direction to the boat. Then it begins to angle round, following the arc of the line and slightly increasing in speed (Fig. 22). If rainbows are there they seem quite unable to resist the lure as it is accelerating round the bend. The method can be deadly, but for some reason some anglers do not get results. I think what they must be doing is casting across wind and allowing the line to swing round without recovering as it does so. This would minimise the acceleration round the bend and leave them trailing the fly directly behind the boat. I have shown the correct method to a great many anglers now, and they are invariably impressed.

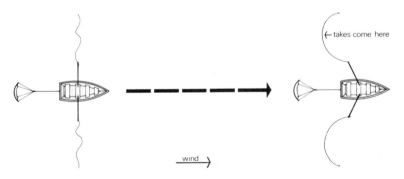

Figure 22

How the line bows after a crosswind cast. The movement of the boat, with the wind, causes the line to curve and the fly to arc round with it, eventually to a position dead astern of the rod tip. But the trout usually take before it gets that far, mostly when the fly is accelerating round the curve.

Farmoor II

Like the Farmoor previously mentioned, this is another concrete bowl water which somehow manages to grow good fish despite the absence of natural banks. Decaying algae forms a soft bottom on top of the concrete, and there's a sort of alkaline reaction from the concrete which helps to create a suitable habitat for many nymphs, aquatic insects and crustaceans. Long casting with long leaders, 5 to 6 yards, is the method here, with lures usually getting a good response. But Farmoor II is notable in my mind for a slightly offbeat tactic, which seems to increase catches considerably. After casting, and waiting for the flies to sink deep, walk down the bank five yards before starting the recovery. Obviously you cover more ground, which is likely to help you contact the fish, but the short walk also seems to change the recovery direction from straight line to more of a curve. One way or the other, the fish seem to go for it.

Hanningfield

I'm here at the invitation of David Allen, who is keen to see if Grafham tactics work at Hanningfield. So am I, and I'm equally interested in finding out whether a slightly faster drift with a new miniature drogue will catch rainbows.

I'm in luck because the conditions are such that a rudder-controlled boat would be too fast and a big drogue too slow. The mini drogue gives us a drift a fair bit faster than normal, and we soon catch rainbows on a slow-sinking shooting head. Later we catch good quality browns with the same method, and I'm intrigued to find that the browns are apparently as keen on daphnia as the rainbows. Eventually I discover this is not a case of Hanningfield browns having different tastes from their Grafham brothers but it is simply that browns figure more prominently in the Hannningfield stocking policy. Some 25 per cent of the total stock of 40 000 are brownies and, much as I love rainbows, I wish other waters would think more about long term results and place more emphasis on browns.

Hanningfield (12 months later)

This time I'm the guest of my old mate, Jim Gibbinson, a new but enthusiastic convert to fly fishing after years as an outstanding coarse fisherman. He, too, wants to see Grafham tactics in action, but I'm thinking about the trip a year ago. If good browns would fall for rainbow tactics what might I catch if I fished for big browns with fast-sinking tackle in deeper water? Very slow drifting in 35 to 40 ft of water near the valve tower wall produces some corking brownies which fancied an Appetiser, the best just under 4 lb.

The lesson of this short story, of course, is always to try to fashion your approach to what you know about a water. Fish for what you know is there in a manner which you know will work. It sounds obvious, but the previous year I had been guilty of undertaking a 200-mile round trip and fishing for rainbows, which I could have caught just down the road.

Ringstead Grange

This water was saved from oblivion by farmer Harold Foster, a former customer who got upset when the previous managers gave up on it. Restocking never helped, because a thriving pike population swiped them, so Harold pumped it almost dry and took the pike out, many double figure fish to 20 lb. Now it's a fine asset to anyone who chooses to fish this 36-acre water, near Thrapston, in Northants. I once broke the rainbow record here, with a fish of 6½ lb, but it has gone several times since. I took my fish on a Christmas Tree lure, black, but with a fluorescent red tail and green thorax. I chose that lure because I thought it would show well in the water, which was fairly coloured at the time, but normally it's a good nymph water, using floating lines and long sunken leaders or, better still, a sinktip.

Draycote

Teams of wet and dry flies designed to catch rising fish are singularly unproductive until a companion offers a size 12 dry sedge on an ungreased leader. We go on to take a limit that way, after hours of failure with a greased leader. One object of greasing a leader is to create a small wake by dragging the fly. At times this attracts trout, but on this particular day it is too calm. A No. 7 weight forward floater, sunken leader and one dry fly, sitting high and not moved at all, amounts to perfect presentation.

There are two alternatives which have proved their worth on many waters. I often fish with a dry fly on the top dropper with two nymphs below. The leader is greased to the top dropper only, which allows the nymph to fish in mid-water. I catch on the dry sometimes, but if not it acts like a coarse fisherman's float in giving early warning that one of the nymphs has been taken. In calm conditions it is retrieved slowly but in a breeze I merely allow the breeze to fish the flies for me. The other alternative I believe I have mentioned earlier – leader greased right along to a dry fly on the point with two nymphs above it, hanging in the surface film on ungreased droppers.

Blithfield

This Staffordshire reservoir is a large season permit water, which I have been delighted to fish from time to time as a guest of members. Some of my visits have produced problems to over-come, and there may be lessons for readers in the way it was done. One time the water was really murky, thick with an algal bloom. Experience at Grafham and Chew in such conditions told me a sunken line and lure approach would fail, so I chose a floater with three small wet flies. One fish fell to a hairwing Butcher. A cold strong wind was blowing, and I decided to change the fly team, leaving the Butcher on the point but changing to a Baby Doll on the middle dropper and a bushy, palmered Sedge on the top. Almost immediately the Baby Doll was taken, and we went on to take a total of 10 rainbows when the wind eased a bit. A wind lane formed, and we caught most of the fish on the Baby Doll, perhaps because the bright white colour made it more visible.

The next trip was a marked contrast, for the water was clear and hot sun was blazing down for the fourth day running. Nobody had caught for three days; not promising, but the message from previous big water experience was to fish deep with the fry imitating Appetiser, using fast sinking shooting head and 5-yard leader. The very first drift produced four fish, and we went on to take a limit of ten apiece, shaking off as many again. Not bad for a day when they weren't supposed to be feeding But imagine if we had reversed tactics, on those two trips, fishing deep in the murk, and up top in the heatwave. I doubt that we would have caught a fish.

Yet another Blithfield trip produced little information of tactical value, but there was an example of the voracious nature of rainbows in warm conditions. I caught one which had inside it a perch a third as long as itself, and it was still stuffing itself with buzzer nymphs.

Eyebrook

Bank fishing and nothing doing until a terrific wind gets up from left to right. Waves start churning up the bottom in shallow water just beyond the point I'm fishing. Casting just off the edge of the discoloured water produces fish after fish, and the experience paves the way for many such catches in the years ahead.

Another tactic for fishing from a point, which is not peculiar to Eyebrook but which seems to fit in well at this juncture, concerns the use of algae. It can be employed, at certain times of the year, though mainly in May, and it amounts to a dodge that some people might consider a touch irregular. When the weather begins to grow warmer algae mats in the shallows break up and drift to the surface. Often the breeze forms the algae into long lanes and these prove attractive to the fish for two reasons: there is food in the algae itself and in the slack at the lee side of the algae lane. This is a tiny strip of calm in which insect life automatically gathers, and because it is calm trout cruising below can see the food more easily.

Fishing calm lanes is a recognised tactic, but in a cross wind at a point, or in a back wind anywhere on the bank, you can deliberately produce an algae lane by scraping the bottom with your waders. The algae lifts up and a lane gradually forms and extends. Fish eventually find it and home in on you up wind. I have had them curving in at me along an algae lane from 200 yards out, and they eventually get within casting distance. Usually the most successful approach is a floating line and a team of nymphs, but traditional wet patterns such as Greenwells, Claret & Mallard, Blae & Black and Wickhams are also very effective.

Draycote

Force 8 wind blowing and nobody seems to know whether to go out at all in the boat or to seek some inshore shelter. Hard work on two sets of oars sees us battling up wind, aiming to fish a drift past Lin Croft point towards Croft Shoal. A drogue proves inadequate so we throw a big three-pronged hook over the side, attached to heavy chain and a rope. It succeeds in slowing us down and then snags fast in a real jammy spot. We are stuck in a perfect position to fish a 20 ft channel midway between the point and the shoal, and take brownie limits on the lead line and Black Lures, fished slowly along the bottom. The whole time we are tossed by waves three to four feet high, but who cares? Draycote boats are sound and the gale is forcing water, laden with food, through that channel. We abandon the hook, tying on a floating marker, and repeat the performance next day by tying up to it again.

Grafham

Now for the third lesson that the wind is your friend, not an enemy. A gale is pounding waves against the dam wall, stirring up the water and sending back a brown undercurrent the opposite way. We inch towards the dam by raising and lowering the anchor, looking for the spot where clear and coloured water meet. We find it 200 yards out and have a birthday; the boats which sought shelter have nothing much to report.

Ardleigh

Not much of a lesson in this trip. I just want an excuse to mention the water which gave East Anglia its first real taste of reservoir trouting. The place is a real credit to fishery manager, Richard Connell. We can't stop catching stockies, though, until we opt to fish deep with fast-sinking lines. This brings some good early season brownies, a bit thin but a good pointer for later on.

A later trip partly proved the brown trout potential, but not conclusively enough for my liking. Driven to fish deep after catching a lot of stockies, my White Marabou Muddler is crunched by a big fish which hugs the bottom, circling doggedly. The line does the old familiar cheesewire job on the surface of Ardleigh, but I am destined not to learn how big the fish is, for after two or three minutes, in which I gain no line at all, the hook pulls away.

Mick Nicholls, Bob Church's regular boat partner, nets a nice brownie from the East Anglian reservoir, Ardleigh.

Good sport up top later, on the evening rise, and I return with my first impression of Richard Connell's work amply reinforced. He has since pulled a quite ingenious stunt, opening up Ardleigh to all methods from October, knowing his season's stock was pretty well depleted. In that spell bait anglers had roach to 3 lb 6 oz, and some very good perch, which have been taken here to $4\frac{1}{2}$ lb. The enterprise changed what would have been a slight loss on the season into quite a useful profit. Other waters could exploit their coarse fishing better than they do – notably pike at Grafham and Rutland – to the mutual benefit of anglers and fishery balance sheets.

Avington

This much discussed Hampshire fishery has changed the face of the record list, and may eventually do the same for the entire sport of fly fishing. Where else can you catch a record rainbow and umpteen double figure fish in a day and get nothing but criticism for it? That is what happened to a party of us back in 1977. Alan Pearson got stuck into an 18 lb 7 oz rainbow, then the biggest ever caught anywhere. I had one of $15\frac{1}{2}$ lb, and can still remember the awesome sight of an entire fly line in the air between rod tip and leader as my monster hit my size 10 Ace of Spades offered on a 7 lb leader. Dick Walker, who had taken an 18-pounder there a year earlier, and Peter Dobbs, also connected with doubles. All good fun, if you don't take it too seriously, but some people did, with letters to the newspapers criticising the alleged gimmick of Sam Holland, the owner, in releasing a few giants for big name anglers to catch. Whether that is what he did I don't know for sure, but I'd be surprised if he didn't, for he has a keen eye for publicity. He professed himself to be disappointed, though, because nobody had caught his bigger fish, which he reckoned went to 30 lb!

I used to get steamed up about instant monsters myself, and there were some pretty bizarre happenings, including a young lady breaking the rainbow record on her first trip. But now I accept these fish for what they really are – swimming advertisements for Sam Holland's extraordinary breedings techniques. It is true that he has somewhat distorted the record situation, but

Alan Pearson with the record brook trout, $5\frac{1}{2}$ lbs. He also holds the rainbow record.

we can all retain our own opinions as to the relative merits of anglers' achievements. For me the best rainbow catch since Colonel Creagh-Scott's 8½-pounder from Blagdon in 1924 was Fred Edgeworth's 9-pounder from Grafham in 1975. It undoubtedly grew on from introduction as a mere stockie, and must have evaded a lot of flies and lures before Fred caught it. I have had several bigger rainbows than that one, but I'm not kidding myself that I have done anything special. The main thing anglers must realise is that if clever breeding can achieve faster growing strains of rainbow the benefits are likely to be felt much further afield than Avington.

Sam Holland has gone further than just messing around with the rainbow, which was itself derived from the American steelhead. He has grown another transatlantic import, the brook trout, to hitherto undreamed of sizes in Britain. His strain of disease resistant rainbows have also been further crossed with Artic Char, producing sterile mule fish which will grow to around 4½ lb in two years. Rainbow × brook trout have produced the aptly named Cheetah, which is built for speed and the product of brook trout × brown trout has been called the Tiger, again aptly named, for it has white bars instead of spots. The point about all these 'mules', apart from the obvious fact that they add variety, is that, being incapable of breeding, they remain in tip-top condition for 12 months of the year, opening up the possibility of all-the-year-round fishing. There is little point in a close season for fish which do not breed, and many have felt for some time that it is pointless to protect the rainbow, or even the brown trout in some stillwaters. Rainbows, depending on the strain, achieve spawning condition in spring or autumn, but cannot breed naturally. The browns can if they can get up feeder streams, but there is little logic to the close season for most stillwater trout. I am sure there will be a gradual expansion of 12-month seasons, which already apply on a few small waters. I don't see this sort of fishing achieving mass popularity (it's hard enough on the fingers taking in line in the spring!) but there is surely a flimsy argument against longer seasons, which could greatly enhance the viability of many waters.

All this, of course, has precious little to do with how to catch trout, which is the overall aim of this and the preceding two chapters, but I felt it had to be said. There is little point in my

Avington rainbows weighing $15\frac{1}{2}$ lbs and $11\frac{1}{2}$ lbs taken by Bob Church and Pete Dobbs.

Bob Church playing and landing one of Avington's giant rainbows . . . an experience Bob has clearly enjoyed!

going extensively into the business of how to fool the monsters of Avington (it's claimed there is nothing to it anyway) but I can cover it briefly. The main enjoyment is in trying to sort out the bigger fish, and in avoiding the smaller ones. It's done by sitting and watching for a biggie to show, and casting accurately into its path, usually with a leaded nymph such as the Dick Walker patterns, the Damsel Fly green nymph, Mayfly nymph and the Westward Bug. Working the nymph at the same depth the trout is swimming, and in as close as possible to the same direction, is often a deadly ploy, but more standard nymph fishing methods also succeed, especially allowing the nymph to sink and then offering it to the trout with a precisely timed pull, which causes the nymph to rise smartly into view. The former method will account for a fish which isn't actively feeding, giving it a longer look and more time to make up its mind; the latter can trigger an instinctive grab from a fish in the same mood or trigger a reaction from an actively feeding trout.

Chapter seven

Lochs and the loch style

To many modern British anglers stillwater trout fishing has a very short history, dating back to Grafham's opening in the early 1960s. That supremely happy event, however, merely brought the sport into the easy reach of the masses, and how eagerly we grabbed the chance. Previously such sport had been available only to those who could travel to the big loughs of Ireland, and the lochs of Scotland, and they tended to be the small minority interested in a change from river fly fishing. Once Grafham had taught us what we had been missing, more of us began to look outward, investigating the waters where our sport really originated, centuries ago. For me that experience has been uplifting and, in one sense at least, depressing. It's all there, but just too far out of reach for me to spend more than a fraction of my remaining time in such a paradise. One look at Ireland was enough for me. I could see why it brought the Americans across the Atlantic in droves, and I only had to cross a relative puddle to get there. If I could spend only a fraction of my time there, I was going to make sure it was as big a fraction as possible! More or less the same thought process led me to the same conclusion about Scotland, and when involvement with the England fly fishing team forced my attention back to traditional loch fishing methods my life was irrevocably changed. If I had to learn how to make the loch style work here, how could I avoid trying it where it all began?

It was not the loch style itself which first drew me to Lough Mask, Co. Mayo, but the experience of Fred Wagstaffe, who once had three trout between 10 and 11 lbs in one week, followed by a 13½-pounder later. In the same period the late Bill Keal had one over 16 lb, and what trout angler could resist that? I went to fish the method which had succeeded for him, trolling the 30 ft contour with the lead line, and an echo sounder to keep us on course. Mask, an awesome 22 000 acres which Fred had christened 'the world's number 1 trout pond', unfortunately did not want us to fish that way. It displayed one of its wilder moods, confining us to more standard forms of fly fishing in sheltered bays. The frustration of our main ambition did nothing, though, to prevent us enjoying ourselves immensely. The conversation alone was a joy, for we stayed with Robbie O'Grady, at Ballinrobe, who fishes the lough every day of the season. His record is formidable – over 50 double figure brown trout in a dozen years, and maybe a few more since our visit. His expertise is such that he is the only man ever to win the World Cup Wet Fly Open Championship twice, a competition which involves hundreds of anglers from America, Canada and Europe. He set fire to our imaginations with his stories, reaching back to his teenage days, when he had once hooked a monster trout while dapping with three Mayflies. It leaped several times before he lost it, a full mile away from where it had been hooked, and saw enough to estimate the weight at around 17 lb. That's not out of line with what Mask is known to have produced. Catches have included an $18\frac{3}{4}$-pounder on bank fished lobworm, and one just short of 20 lb by a French lady in 1977. Neither of those fish was considered old enough to be one of Mask's real veterans.

On our trip we learned to appreciate how it must have been for the young Robbie O'Grady, when the big fish took his Mayflies. The smaller fish we caught took to the air like rockets, wild, golden-flanked trout which had never seen a hatchery pellet. They would not take our lures, for some reason, though Robbie felt they might work well when the trout were having their annual gorge on the perch fry. But they would have a go at a Mayfly nymph, fished among the rocks (we were too early for the annual Mayfly hatch) and they responded well to the broadside on drifting style, to be described later. The killing flies, we found, were Teal & Black, Mallard & Claret, Invicta and the Irish favourite, Sooty Olive, and several times we saw trout arrowing up to take them, from several feet down. We left Mask with great regret, but determined to return. Where I shall find the time to fish other notable Irish waters I cannot imagine, but the call of Corrib cannot long be ignored. And there's Conn, Derg, Arrow, Sheelin, Derravarragh, Caragh, Owel . . .

A double-figure brownie from Lough Conn.

Scotland offers the same bewildering choice. In Caithness alone there are 110 different lochs, many of which are never fished. I fished two of them with Thurso tackle dealer Sandy Harper, and thoroughly enjoyed it. Watten, the biggest loch in that area, three miles long by a mile wide, produced splendid brownies averaging around 1 lb, but I also enjoyed Loch Caol, a smaller water on Lord Thurso's estate. It can be reached only by a long hike, when the car runs out of road, but the experience was tremendous. The fishing was good, but the main memory is the wild beauty of the place, and the utter silence. I cannot mention all the lochs and loughs where I have wet a line, but there is a common thread which seems to link them all. The methods we have learned for the English reservoirs can work well, and I have yet to meet an Irish or Scottish angler who has not had his horizons slightly widened. Some of them are clever anglers, and therefore open minded. They have seen that they can usefully borrow some of our 'technology,' and we have certainly borrowed some of theirs. The loch style varies hardly at all between Scotland and Ireland; only the fly patterns used are different, reflecting to some extent the local conditions. Learning how it is done gives us more catching power at home, and it certainly paves the way for much enjoyment on holidays. At first glance there is nothing to it, but nobody who shared a boat with the late Cyril Inwood believes that. He outfished everybody, and it is only now that I fully understand why.

The method itself is simple; short, delicate casting from a drifting boat, and with a team of three flies. Do it well and you can empty the place; do it badly, though, and you start to doubt the efficiency of the stocking policy. The bob fly is the key, the one on the top dropper. It is fished in a way which attracts the fishes' attention, and they either take that one or, sometimes as they turn away from the bob, one of the others below. I'll take you through the routine for a day when there's quite a big wind and big waves. The bob fly, as always, is a bushy type, and the bigger the wind the bigger the fly should be. A typical threesome would be a Soldier Palmer on top, a Mallard & Claret in the middle and a Butcher on the point, but I'll mention other possibilities later. The cast in rough water is very short, 5 to 8 yards at the most, and the boat is very likely drogue controlled for drifting in such conditions, set so the drift is broadside on. The aim is to dibble the bob fly in the waves, so that it creates a bit of a wake, and casting is constant – show 'em the flies, whip them off and back again. With such a short line it has to be that way, for the boat is drifting towards the flies. They have to be recovered faster than the boat is drifting, keeping a tight line to the flies the whole time. It sounds a bit frantic, but it's not; more a gentle, rhythmic process which I find quite relaxing, until a silvery thunderbolt erupts to the top and grabs one of the offerings.

The scenario changes a little when the winds are lighter, and the waves smaller. We find, then, that the fish cannot be fooled quite so near the boat. The cast now is around 15 yards, and in order to work the flies in the same way we need the longer rod referred to earlier, $10\frac{1}{2}$ to 12 ft. It gives us the same control at longer range for dibbling that bob fly in the waves, only now that fly is dressed on a smaller hook, a 14 or 12, instead of a 10 or an 8. When the wave action is small, however, the trout will not always respond to mere dibbling, and when they don't, a change of tactics is called for. We cast that 15 yards and recover it in four or five smart pulls, holding the rod almost parallel to the water instead of quite high, as the dibbling approach demands. Only when the flies are near the boat is the rod raised, and the flies dibbled on the surface for a few seconds before the re-cast. That's to nobble the trout which has seen the fast stripped team but hasn't quite had the legs to catch them. It often falls for the last ditch dibble, and many's the time I've seen them fall like this to Inwood, who seemed to get them almost jumping into the boat for his offerings. There are times when a couple of fast strips and a dibble will catch trout in the big waves, but generally speaking big waves mean dibbling with a big bob fly, and lesser waves respond either to that or the fast strip. I don't suppose anyone reading this will think they could not easily master this technique, and they would be quite correct. But it would not put them on a par with the likes of Inwood, or some of the current internationals for, as usual, there is more to it than meets the eye. The key to success with this method is absolute concentration, and there's more than the flies to look at.

For at least part of the time, each cast, you should be looking for the fish, scanning the area just coming into casting range. If you see one moving upwind – and the sign can be very slight, often no more than a smooth patch of water in a wave – try to calculate where it is likely to be by the time you can cast, and drop the flies right there. It isn't as difficult as it sounds. The likely line the fish is taking is dead upwind of where you saw it, and when you are working with such a short line it takes but a second or so to whip up the flies and drop them on its nose. If you fall short the fish is likely to see the flies, and put on a bit of a spurt to catch them, and if you overdo the cast you are still likely to pick off one which is following on behind, for rainbows travel in packs. Even when there are no surface signs, the chances of catching with this method are high, providing the conditions are right. It works in the warmer months of the season, when air temperatures are normal or above. An ideal wave height is 6 to 9 inches, but anything up to $1\frac{1}{2}$ feet is good. Given these conditions the fish will be high in the water, even if they aren't showing themselves, probably cruising upwind, a foot or so under, constantly on the lookout for food such as daphnia, small nymphs and so on. These high fish are actively hunting for food, and very vulnerable.

The method has a double edge, with attractor flies working in the rougher conditions, and imitative patterns more likely to succeed in calmer conditions. I vary my leaders according to the conditions, too, using two yards of 6 lb line, one yard of 5 lb and one yard of 4 lb for the calmer water, stepping up to $1\frac{1}{2}$ yards of 7 lb, a yard of 6 lb and a yard of 5 lb when it is rough. It's a better match for the larger flies, and is better able to stand the extra strain rainbows can exert in the bigger waves. As I said earlier, the key to success is choosing the right bob fly for the conditions, and there's quite a long list of suitable candidates. Some of my favourites are the Irish Murrough, Thicket, Red Tag, Invicta, Soldier Palmer, Orange Sedge and an American fly which has proved a good one, Queen of the Water. Flasher patterns which are a long way ahead of the rest are Dunkeld and the Butcher family, with the Bloody and Kingfisher Butcher superb in the warmer conditions. Among the best deceiver patterns are Greenwell's Glory, Blae & Black, Ginger Quill and Olive Quill, together with the previously mentioned Invicta and Claret & Mallard. There's plenty of scope for experiment, but after a while most anglers seem to settle for teams of flies they have confidence in. The trio might change with the conditions, of course, but after a bit of experience the right three will suggest themselves.

There are one or two tricks in the loch style trade, mostly connected with the way in which the boat is used. Preferably, it's a method for two good friends, working as a team, each one taking half the water ahead, and neither encroaching on the other's territory. They can help each other with fish spotting, for we often see fish which are best covered by the other angler, and by using a crisp sort of code we can pass the information fast enough for the companion to react. Some of us use the old RAF code – 'bandits two o'clock high' stuff. Given a straight broadside drift down wind 12 o'clock is the point directly ahead, between the two anglers. The message 'two o'clock twelve' takes only a second to transmit, and in another second or so the team of flies can be dropped on the spot, 12 yards out. It's great when it works; an extra thrill for both anglers, but you can find yourself in a boat with someone less cooperative, and not averse to a bit of gamesmanship. For example, it is possible to alter the drift direction by leaving the outboard motor in the water, instead of on the tilt clip. The motor then acts as a rudder, and instead of

Thurso tackle dealer Sandy Harper (*left*) and Bob Church with a Loch Watton brownie catch.

drifting square on to the wind the boat tends to move away from the angler sitting in the stern. His flies are then being offered over water not covered by the drifting boat, and he can work his flies right back to the boat – a major advantage, especially in high wind and wave conditions. The other man in the bow can only move his flies to the centre of the boat and has to lift off much earlier, because the boat is sweeping down on his flies. If somebody does that to you, insist that the motor is lifted! But then there's half a chance of him turning the tables on you, especially if he is a heavyweight. Crafty beggars like Frank Cutler – who is particularly heavy – know that if they sit as far towards the bow as they can get their weight will make the bow bite much deeper into the water, where it acts in similar fashion to a leeboard. It alters the drift in such a way that he can just cast out without retrieving his flies. The boat moves them for him, and he just works the rod so that the flies come back to the boat in a deadly curve. These little points are worth remembering for competitive fishing, but when fishing with friends the fairest approach is to change positions every hour or so.

Dapping

Dapping is a method which is very much an Irish lough approach. For some reason it has not travelled well, and I can't understand why, for it works well enough on all sorts of waters. Maybe it has stayed in Ireland because it is mostly associated with the big Mayfly hatches they have over there in early June. Our Mayfly hatches are less impressive, and confined nowadays to relatively few waters, but the Irish don't just dap with Mayflies. When that brief bonanza is over they turn to the large Sedges and grasshoppers, before finishing their season with the Daddy Longlegs. In our conditions I feel the Daddy Longlegs is the best bet, though grasshoppers should work well too. I haven't had much luck with them, though, for whenever I have gone to the trouble of collecting them the conditions for the method have turned out wrong. I'll try again, but have most faith in the 'Daddy', having seen it account for some big fish and big bags at both Grafham and Rutland in recent autumns. Tackle required is a carbon fibre rod of at least 12 ft, but 14 to 15 ft is better. The line is a special floss blow line, thick but very light. It catches the wind well, and you work with up to 25 yards of it, attached to the usual backing. For a leader you need two yards of whatever breaking strain is considered appropriate for the fish you expect to catch, but for me it is normally 6 to 7 lb. We have got the art of tying Daddy Longlegs artificials to the point where it is hard to distinguish them from the naturals, but make sure they are well sprayed with silicone floatant. If you wish to fish the natural insect, though, you need size 12 and 10 fine wire hooks, impaling two on the hook. With grasshoppers the hook has to be carefully passed through the back, and they should be kept in a narrow necked bottle, with airholes in the top, so you can extract just one at a time.

It's possible to dap on any water if the wind is right, and it can be great fun on rivers for trout and coarse fish. I prefer it, though, from boats on the big waters, drifting broadside on, as with the traditional loch style. The rod is held almost vertical as 15 yards or so of line is fed out. The wind will hold it well above the surface as you feed the line, but when the correct amount is out you then lower the rod so that the offering, natural or artificial, touches the surface. Jigging the rod top makes the bait skitter on the surface in a series of little jumps, and you cannot achieve this with any of the standard fly lines. When it is working well it provokes takes which are best described as violent. You have to school yourself not to respond in similar fashion, or you will either pull the bait away before the trout has got it properly, or, worse still, break on the strike. Give the fish a second or two to get hold – it might miss the first time – and gently lift into it as it is heading back down. It isn't a method which is going to sweep the country, but it can be a very enjoyable way to spend a few days at the end of a long season.

Chapter eight

Methods in a nutshell

Looking back over the previous chapters makes me realise that things which I take for granted might appear complex when written down, at least to the beginner. More experienced anglers will have to forgive me, therefore, if I attempt to summarise. I will try to establish what the correct basic approach should be in given conditions.

Early season (Cold water)

Almost invariably the successful approach is to fish slow and deep, either with lures or nymphs. Choice of line depends a great deal on the depth – floaters for shallow water, sinktips or slow sinking shooting heads for deeper water and fast sinkers for depths over 20 ft.

Early season (Warmer water)

Slow and deep still likely to be the best, but lures can be speeded up a little. Nymphs are always fished slow, whatever the time of year. Possibility of catching in mid-water or up top with nymphs, particularly if there is a buzzer hatch.

Mid season (Warm and calm)

Slow and deep with nymphs, or floating fly with nymphs hanging in mid-water the likely methods, especially if nymphs and dry flies imitate what is hatching naturally. Lures fast-stripped across the surface occasionally work, but lures fished slower over the bottom are likely to work even better.

Mid season (Warm and windy)

Sinktip line and flashy lures (especially orange and fluorescent lime green) will take rainbows fairly high in the water if they are feeding on daphnia. The orange and green can be used as spots at throat and tail, or as full-size coloration. Lure recovery fast. More sombre lures fished slower and deeper will catch if daphnia and/or algae is not present in the upper water layers. Dry fly and nymphs effective during hatches, as is loch style drifting in good winds.

Summer and autumn

All mid-season approaches will work at times, together with lure fishing for the fry feeders in the manner previously described.

Fishing in the wind

Seeking to get the wind directly behind to assist casting is usually a mistake. Fishing across or into the wind almost always gets the best results, except during a hatch of sedge or buzzer. The best hatches often come in the calmest water where the wind is off the back or, of course, when there is no wind at all. If there is nothing hatching, avoid the 'flat' water. Boat anglers should take the same advice as far as the calm areas are concerned, but their main approach should be to fish areas where the wind is creating currents and washing food to waiting trout.

It also occurs to me that I have buried a great deal of information about the location of trout in the descriptions of my fishing trips, so perhaps I ought to summarize that as well. Location is, of course, no problem when the fish are feeding up top. You can see them rising, and by watching the water carefully you can see when they are feeding just under the surface. The water humps and boils as they grab a food item and turn, but do not expect the signs to be obvious all the time. A tiny sip on the surface and spreading rings can be fry taking the air. But it can also be a 5 lb trout on the feed. A smooth patch in a wave can be caused by a trout, so keep your eyes skinned. The only problem with surface and sub-surface feeders is identifying what they are feeding on and offering an acceptable imitation in an efficient, unobtrusive manner. Finding fish when they are not showing is another matter altogether, and in effect a large part of this book is about locating them. Grafham, for example, is six square miles, and the rainbows could be anywhere. I have told you how daphnia influences their whereabouts, and I expand on that later on. But if conditions – notably wind and light – do not help us to make an intelligent guess as to where the rainbows are, we go looking for them by covering miles of water with one or other of the drifting techniques. Fishing at anchor, drifting down wind 50 yards and anchoring again is another method of searching, albeit much slower. Walking the banks is slower still, but it has to be done if you have no idea where to find fish. I have mentioned all the other ways to locate fish by identifying areas where there is a good chance that they will be, but there is no harm in repeating them in summary form, so here goes:

Dam walls

Rocky walls such as that at Draycote hold all manner of fish and insect life, and trout are never very far away in any conditions. Any dam wall onto which the wind is blowing offers possibilities, because the wave action washes food back in the undertow to waiting trout. You can catch those trout by casting into the wind from the dam, or anchoring 50 yards or so from the dam and fishing towards it. This method is particularly productive if the wind is strong enough to colour the water. You anchor where clear and coloured water meets.

Weedbeds

Always likely to produce trout, sometimes very big ones, when the fish are after fry. The fry hug the weedbed for protection, but it doesn't offer much when the trout have a taste for fish.

Hedgerows

A number of reservoirs cover flooded fields bounded by hedges. You can often tell where the submerged hedges and ditches are by following the line of hedges and ditches running across the reservoir banks to the water's edge. In many cases the remains of the hedges and the scoop of the ditches are still there. Such areas collect fry and other food, and are happy hunting grounds for trout and anglers.

Inlet towers

These or any other reservoir furniture which influences water movement attract small fish – and trout.

Algae lanes

The wind pushes algae into lanes, and food items gather in the tiny crease of slack formed by the wind on the leeward side of the lane. A feeding trout will sometimes top and tail its way hundreds of yards along an algae lane.

Wind lanes

These are often much broader than algae lanes. They are strips of flat calm water running across a choppy surface, and they are caused by natural or man-made features on the reservoir bank. Electricity pylons, church steeples and high hills break up the wind, an action reflected by flat areas on the water. Food naturally collects in these lanes, and so do large numbers of trout.

Points in a crosswind

A small peninsula of land is often a likely spot, especially when the wind blows across it. The wave action colours the water and stirs up the food.

That's the skeleton upon which the general approach is based. Anglers who stick to that, but are prepared to make intelligent changes based on observation of what is happening on the day, or during part of the day, will get the best results.

Boat or bank?

The discerning reader will no doubt have noticed the fact that the preceding passages, whilst covering bank techniques fairly often, show a definite bias towards boat fishing. I must go into this more deeply at this stage, after first making the point that every method mentioned in the context of boat fishing is equally applicable to the bank *except* where the success or otherwise of the technique depends in some way upon the manner in which the boat is used. In short, if the boat is merely a floating platform from which to fish the method employed will also be relevant to bank fishing, given the same set of circumstances. In addition *some* of the moving boat techniques are applicable to bank fishing. One reason for the drift technique, for instance, is to find fish. The bank angler's equivalent is to use his feet, instead of taking root all day in a spot which may never produce. Another example is when the movement of a boat influences the speed and manner in which a fly is fished. A cast across wind from a moving boat puts a bend in a sinktip line and, as you have read previously, the recovery through a curve finishes with a final whip round of the fly which often provokes a take. The boat angler cannot *avoid* that bow even if he wants to. The bank angler can *allow* a cross wind to bow the line before retrieving. The speed of a boat obviously adds to the speed of recovery, calling for a slower retrieve by the angler himself. The bank angler, by pulling in faster, can exactly match the presentation. What he does *not* get, however, is the ability to cover vast areas of water as efficiently as the boat angler, and he also misses something vital – the opportunity to fish all day where the fish are.

My approach should be clear enough by now. I always want to be fishing the likeliest method to catch fish and that, more often than not, suggests the use of a boat. On most waters early in the season it is not just possible to catch fish from the banks; it is also easy. Towards the end of the season, particularly right at the end, bank anglers are more likely than the boat men to catch big fish. The fry move in to populate weedbeds growing in the margins and the trout follow, thus exposing themselves to the bank man. Throughout the season trout can be caught from the bank in circumstances mentioned in preceding chapters. But for long spells in mid season the amount of time the trout are vulnerable to the bank man is short – too short for my liking.

We all know about early morning, when the trout are in after a night on the shallows, and about the evening rise, which may or may not happen. These are vital times – almost magical moments, sometimes, when we bang the hook into an early morning fry raider or deceive, at twilight, the cautious monster which lost all inhibitions when it first caught sight of a fluttering, struggling cloud of sedge. But moments like these, magic though they are, are mere book-ends, or brackets which enclose an entire angling day. In between, when the water in the shallows is clear and the trout are out in the wide blue yonder, the best fly angler in the world is in trouble. Very often his only hope is to spend the day scraping the bottom in deep water off the dam wall. At times this can produce brownies big enough to send him back to the fishing lodge with chest expanded, but most of the time it won't.

On days when everything is against the bank angler there are not even any brackets to his day. He draws a blank, through no real fault of his own. Total blanks, however, are fairly rare for the accomplished angler. He can usually winkle a brace from nothing, usually by plugging away with an imitation of something he knows will be present, since difficult days are not the days when fish are likely to respond to an attractor. An odd fish might stray into range and be caught with a neat, accurate cast, or the angler may find a couple by wandering off to the more distant and inaccessible areas where tramping feet have not scared trout to the horizon. There is extra satisfaction in the 'struggle fish', and they tend to hallmark the successful angler as a cut above the average. Some anglers, indeed, are prepared to settle for this as the norm, and regard the occasional triumph at dawn, dusk and odd times in between as the icing on a delicious cake. I could never plump for this approach, though I do respect other people's right to regard it as the be-all and end-all of fly fishing, if they so desire. It is an attitude, however, which is peculiar to fly fishing and it offends against the first rule of fishing. In all other branches of our sport the key to consistent success is to fish where the fish are. You would not catch a member of a coarse-fishing specimen group fishing an empty swim into which whatever he was after *might* stray, especially when he knew he could go to another and much more likely location. But this, in effect, is what fly fishermen who spurn the use of boats are often doing.

I ask myself where the fish are likely to be, and if it is the bank I will fish the bank. Usually this means early in the season, late in the season and early morning and evening in between those times. It should not be forgotten that the boat angler is not denied the opportunity to fish the bank before and after he goes out! It is possible to enjoy the best of both worlds, and I do. Getting afloat offers no guarantee of success, but it does remove one of the factors which influence success. It transfers you from where you *hope* fish will come to where they actually are for extremely long spells throughout a season.

The other major factor which influences me is that a boat is that much more versatile than a patch of ground. All you can do with a patch of ground is stand on it or sit on it. You can do the same in a boat, but you can also make it work for you. I have already expended a few thousand words telling you how. That is my explanation for the boat-fishing bias. They help me to reach fish when they are practically or totally unreachable from the bank, and they add to the tactical armoury. The case for using boats is more often stronger than the case for standing on the bank, so I end up fishing in boats more often. It is all part of my logical, rather than emotional, approach to fishing. I do what *has* to be done, not what I might *prefer* to do. I love bank fishing just as much as boat fishing, and on balance the former is more comfortable. It is also a lot cheaper, but these factors influence me only when the chances from the bank are better or equal to those from a boat. If not I will stand the cramp and pay the necessary.

Chapter nine

The diet of trout

By long experience most of us know what trout eat most of the time, but they are no different from us in the sense that they occasionally enjoy something different. Just as we eat new potatoes and strawberries at the appropriate time of year trout will welcome the occasional change of diet, and when their enjoyment of a particular item becomes a real preoccupation the angler who lacks an acceptable imitation is in serious trouble.

I have previously mentioned two of their occasional snacks – caenis and the now rare mayfly – but there are many others. A classic example, I suppose, is the drone fly which appeared in large numbers at Grafham the year before Cyril Inwood died. He saw what they were taking and went home and tied an artificial so much like the natural it looked as though it could fly. Cyril slaughtered trout at Grafham for three weeks. Casual visitors and the great majority of the regulars did not get a look in until the drones buzzed off. Now we all have drone flies tucked away in our wallets, but they were not seen again in large numbers at Grafham until August 1975, when Cyril's pattern again proved a killer. There are, however, many insects which do put on the odd appearance every season – the daddy longlegs and the flying ant, for instance – and the reservoir regular who does not want to be caught with his pants down should carry imitations of these and other insects. I have included some of the 'occasional flies' in the later chapter about fly patterns, but they should only be used, of course, when observation and/or spooning fish indicates that you should.

The advantages of spooning trout, especially the first one caught, cannot be over-emphasised. It can put you on the right track for the rest of the day, providing you know how to recognise the stomach contents. If you can't then I would suggest you buy a

A marrow spoon full of large food items, which would track the angler straight onto the correct artificial to offer.

suitable reference book, like John Goddard's marvellous *Trout Flies of Stillwater* (A. & C. Black, 1972), which has become my Bible, or Clegg's *Freshwater Life of the British Isles* (Frederick Warne, 1965). Sometimes just one look at the spoon tells you what to do. If it is stuffed with caddis, daphnia fry or snails, for example, the approach should be obvious, if you have followed all that has gone before. The only qualification here, is the case of snails, which I do not believe I have mentioned previously.

Almost all the time trout taking snails pick them up off the bottom, and when they are doing that they seem prepared to grab at any lure, particularly if it is predominantly black. There are, however, odd times when the snails go in for a kind of mass migration, rising from the bottom and drifting with the wind either on or under the surface. It happens in hot weather for reasons we can only guess at. It may have something to do with mating, though the most convincing theory I have read is by

John Goddard in the book just mentioned. He thinks it might be due to slight de-oxygenation which causes them to come up for 'air'. The reason, however, is unimportant. It is the fact which concerns the angler, and when snails are high in the water the trout become very selective indeed.

Our big problem is realising in time what is happening. If the snails are up but under the surface we do not always see either them or the trout taking them, and I can remember one or two dour trips when almost everyone in the fishing lodge was kicking himself at the end of the day for realising too late or not at all. I equally remember an occasion at Draycote when John Snelson seemed to be the only angler catching, and when I asked what he was doing he said: 'They're on the snails.' I looked down into the water and there they were. My special snail imitation (tying later) and a Black & Peacock Spider then accounted for some fish. Any small black fly, heavily dressed with a fat body, is likely to succeed, and on one occasion I caught with a small Black Chenille, fished high and slow.

So, if the spoon reveals lots of snails, fish lures over the bottom with confidence but, if it doesn't work and the weather is hot, take a look at what's under the surface! If the spoon reveals a mass of small items not easily identifiable put the contents in a saucer of water. The various creatures will then separate and you will see which food item predominates, its colour and its size, and you can take it from there. Until quite recently I had never thought about stomach contents in any sort of scientific way. A quick look for an indication of how and what to fish was all I ever did, but that was before I met Viv and Valerie Church (no relation), a husband-and-wife angler/biologist team. Valerie chose the diet of rainbow trout as the subject for an Open University degree course and, I'm happy to say, she qualified on the strength of a scientific study on the diet of trout at Grafham. She very kindly allowed me both to see and report on her findings, and it is most interesting to compare how they square with practical angling experience. The survey covered May to September and involved fifty-eight trout caught by Viv on floating line only. The catches were made evenly over a five month period, and date, time, weather and water temperature were recorded. Examinations usually took place two to four hours after capture. Valerie's findings follow, with a few of my observations.

	Fish under 400 mm							Fish over 400 mm						
	May	June	July	Aug	Sept	Total	%	May	June	July	Aug	Sept	Total	%
Chironomid adults	4	1	2	4	3	14	37.8	3	2	0	1	2	8	38.1
Chironomid pupae	8	4	14	6	4	36	97.3	4	6	3	3	4	20	95.2
Chironomid larvae	6	0	7	1	0	14	37.8	3	5	2	0	0	10	47.6
Caddis pupae (sedges).........	0	1	12	3	1	17	45.9	0	2	3	1	0	6	28.6
Daphnia	4	1	10	4	4	23	62.2	1	3	2	1	2	9	42.9
Ostracods	0	3	6	1	0	10	27.0	1	5	3	0	1	10	47.6
Fish fry.........	0	0	1	1	3	5	13.5	3	1	1	2	4	11	52.4
Numbers of fish caught	8	4	14	7	4	37		4	6	3	3	5	21	

Figure 23
Table showing the percentage frequency of occurrence of the main items of trout diet in large and small fish.

The stomach contents were weighed and then put in water for examination. They were sorted under binocular microscope and identified according to species, using for reference the two books previously mentioned in this chapter and also T. T. Macan's *Guide to Freshwater Invertebrate Animals* (Longman, 1959). Of the 58 fish examined 56 had chironomid (midge) pupae in them – hundreds in some cases and, in others just a few. The pupae varied in size from 3 mm ($\frac{1}{10}$ in) to 20 mm ($\frac{4}{5}$ in). There was a wide variety of colouring, and it was noticeable that numbers of pupae remained alive and active even after being eaten. Some probably even hatched in the stomach. In view of the wide range in size of the chironomid pupae actual numbers bore no relation to amount by volume. In the cases where chironomid pupae was not present one fish was completely empty of food and the other was filled with one large bream. Daphnia was found in huge proportions at the end of July, continuing into August and September. This probably reflects the daphnia bloom well known to fishermen at this time of year, when increased water temperatures bring about great multiplication of these tiny animals. When assessed in numbers the percentage total was 84·57 per cent of daphnia present in the rainbows' stomachs. In September there was an increased occurrence of fish fry in the rainbow diet.

FOOD IN RELATION TO SIZE

The table (Fig. 23) shows the percentage frequency of occurrence of the six main items of diet, plus fish fry, assessed in relation to size of trout. The two categories of size are up to 400 mm (15$\frac{3}{4}$ in) and over 400 mm. When judged in weight, which anglers would normally do, 400 mm equals approximately 1$\frac{3}{4}$ lb in July and 2 lb in August. There is no significant difference in the occurrence of chironomid adults and pupae between the two categories of fish, but the larger fish include chironomid larvae (bloodworm) more frequently in their diet. Ostracods (bottom dwellers) are also more frequently represented in the food of larger trout, while sedge pupae and daphnia feature more often in that of smaller ones. The greatest difference appears in the occurrence of fish fry,

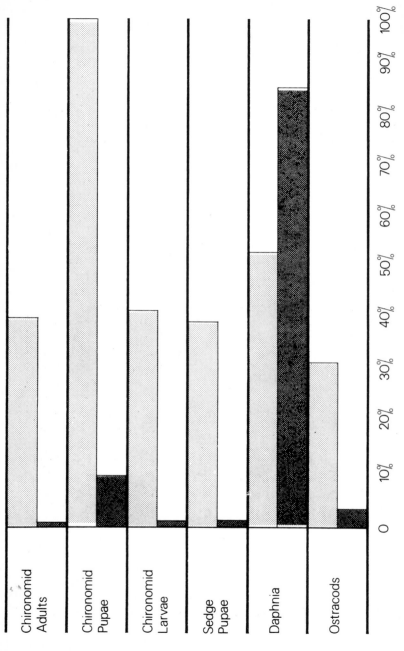

Figure 24

Chart showing the numbers of each main species in the rainbow trout diet expressed as percentages of the total (darkly shaded) compared with percentage frequency of occurrence (lightly shaded).

which was found four times more often in large than in small trout. These were mostly bream and perch. A tentative conclusion from the figures might suggest that the smaller trout feed more frequently from the surface area than the larger trout, since sedge pupae and daphnia occur more often, and chironomid larvae and ostracods less often, in the smaller fish. Chironomid adults and pupae are taken in larger numbers by the smaller fish, though equally often by both categories.

Valerie's findings are, as she took care to point out at the end of her report, based on a relatively small sample, and she thought a twelve-month survey would give a more comprehensive picture. Even so her work confirms many of the findings anglers have arrived at in a much less scientific way. We knew, for instance, of the importance of imitations of midge pupae (see Fig. 24); we knew about daphnia and how it influences catches (more about these foods later); and we obviously knew that big trout do more bottom feeding than smaller fish, and vice-versa. Some of us have also known for years that rainbows eat fry, though many leading authorities have accepted that only recently.

Valerie's work, however, would be very misleading if it was taken as a comprehensive guide to what trout eat in reservoirs. All it illustrates is their basic diet. The frequency of occurrrence of food items, and the amount, would vary a great deal if a similar study was carried out on trout caught by methods other than the floating line. I would wager that there would be significant differences between trout caught on a floater and trout caught on a sinker and, maybe, differences between trout caught on fast and slow sinkers. I know for certain that this statement is at least partially true. A reasonable estimate is that I and my regular boat partner, Mick Nicholls, do at least 50 per cent of our Grafham fishing with sinking lines. The average size of trout caught on the sinker is much higher and, of course, the stomach contents differ considerably. A catch of fourteen rainbows and two browns weighing 41½ lb were all stuffed with 3 to 4 in roach, as were trout taken on several other occasions. I have stressed for years that the likeliest place to catch big trout is on the bottom, and it is nice to see scientific findings supporting one's own experience.

Further studies of trout diet would either teach us a lot or confirm what we know. The only additional information I can offer at this stage is that Viv Church and I continued to look at rainbow stomachs in the October following Valerie's survey, while she was collating the May–September information. Every rainbow except one had fry in its stomach (mostly perch); the one exception was stuffed solid with snails. When almost all surface-caught fish are found to contain fry it tells me that I won't be far wrong to concentrate on fry fishing downstairs if I want the best results.

I am sure there would be other useful pointers to correct tactical approaches if more extensive scientific studies were undertaken on the diet of trout. I would particularly like the same work to be carried out on brown trout.

Chironomidae

Valerie Church's work, together with the experience of anglers, points to the importance of the midge family (Chironomidae) as trout food and, therefore, as trout catchers when successfully imitated. So, let's pause and take a look at the life cycle of what is possibly the most common food item.

The more the angler knows about the behaviour of trout food the more he is likely to catch, for it tells him such important things as where to fish the imitation and how to fish it. The experts tell us there are nearly 400 members of the midge family, so nature has done us something of a favour by making them all behave in exactly the same way. If there were 400 species performing in 400 different ways our job would be a lot harder than it is already. Thankfully the only differences which are important to us are colour and size, and the main colours are red, green, brown and black. Whatever colour they are they all start life in the same way – as eggs laid on the surface of the water by adult females.

Their productivity is fantastic; literally millions are laid every day. After a few days the eggs hatch into larvae, which sink to the safety of the bottom mud and silt. The predominant colour at this stage is blood red – anglers know them as bloodworms – though a good many are also green or brown. The next step is for the larvae to shed a layer of skin and develop into pupae. These remain on the bottom for a while until wing cases form. Experts who have studied chironomidae extensively say the pupa stage lasts for 36 to 72 hours, sometimes longer, and as the time approaches for them to rise to the surface their swimming motion becomes a series of rhythmic kicks. Perhaps you can see why midge pupae imitations catch fish near the bottom with a gently-inched retrieve? They become even more vulnerable to the trout when they do rise to the top, for they do not escape into the air with any speed. I have often looked into the water and seen thousands of pupae just hanging under the surface in the top two feet. I am not sure why they do it, but it is possible that they are either having difficulty penetrating the surface film or they might be waiting for the light to fade before emerging. You can guarantee that when you see pupae in large numbers during the day there will be a massive hatch in the late evening, but you may also see small bursts of activity throughout the day. Maybe they prefer to hatch in the evening but some are so advanced they can't hold back. Whenever the moment comes, however, they break surface, the thorax splits and in just a few seconds the winged insect is up and away.

The angling lesson here is obvious, too. An artificial pupa hanging just under the surface is likely to catch trout, but when there are a great many naturals there the imitation must stand out and attract a trout's attention. It is the tall man who immediately catches the eye when we look at a crowd. It is a big imitation which often becomes the trout's last meal. The fact that the pupae spend so much time hanging under and no more than a few seconds on the surface may also be a satisfactory explanation why we tend to catch mostly just under the surface during a hatch. Most of the trout caught on midge pupae are either taken deep, when the naturals have not left the bottom, or just under the surface. We also catch them in between the extremes, but not quite so often. It is when we are not quite sure where they are that we fish one deep, one at midwater and one up top, but if it is the top imitation which is being taken all the time we might as well fish all three just sub-surface under a siliconed leader.

The best conditions for a major hatch are when the light is fading and there is a softish, warm breeze. On some waters there is a useful early-morning hatch, probably because the light conditions are again right for a spell after dawn. Most of the sporadic daylight hatches occur when it is warm and overcast, though we do see some hatches on quite a cold day. Hatched adults fly to the land, which is why we see great clouds of them on warm evenings. They usually mill about near hedges, trees and buildings close to water. This is when mating takes place, then immediately after, the egg-laying females return to the water to initiate the life cycle all over again. The swarms of midges you see and hear as you fish on a warm evening are the returning females. They seem to like to hang around anglers' heads, and when they do we can hear the distinctive buzzing sound created by a few thousand rapidly beating wings. This is why 'buzzer' has become the collective name for the whole family of chironomidae.

The angler's main task when it becomes obvious that the trout are on the buzzer, is to identify the colour and match the size if the hatch is sparse, remembering to exaggerate the size if the hatch is prolific. Main colours encountered after the larva stage are brown, green and black, but when fishing deep it sometimes pays to fish a red nymph on the point, to imitate the bloodworm.

Sedges

I do not want to turn this book into an exercise in natural history, but after the midge pupae I rate the sedge family as the next most important natural insects worthy of study.

There are knocking on for 200 species in the sedge (Trichoptera) family, and the life cycle is similar to that of the chironomidae – egg, larva, pupa and adult winged fly. With chironomidae we found imitation at the pupae stage to be the most effective angling method. In the case of the sedge it is the larvae which the trout eat most. Regular autopsies lead me to believe that there is no other stillwater larva which trout eat as regularly as the sedge, but this means very little in terms of catching trout.

We can catch them with larva imitations (Stick Fly and Black Chenille), pupa imitations (Brown & Yellow, Brown & Green and Amber Nymphs) and adult fly imitations (Invicta, fished wet or dry, and the Standard Sedge). My results have been about the same with all three stages of imitation, though common sense suggests that if more time was spent scraping around with the Stick Fly and the Black Chenille this method would outscore the others. The larva does, after all, spend twelve months on the bottom compared with just a few minutes up top before hatching and flying away. It is perhaps this prolonged period of vulnerability on the bottom which persuades the grub to build its protective little house. By choosing most of the materials locally available – grains of sand and tiny stones, bits of shell, twigs, rotting vegetation and so on – the grub becomes perfectly camouflaged. I do not know how effective is the disguise in the case of all predators, but it is quite certain that the trout are not fooled. They chomp the larvae, house and all, and when there are large numbers available trout sometimes become totally pre-occupied. I have caught them absolutely choked with sedge larvae from throat to vent, and when they are in this mood they are real suckers for the Stick Fly. Unlike some other cases of preoccupation, however, the need for exact imitation is not great. The Black Chenille is not much of a copy, but it works, and when trout at Ogston Reservoir fed on hardly anything except sedge larvae all summer some years back the most effective lure, I'm told, proved to be a Sweeney Todd.

I have fond memories of sedge fishing on many waters, but I'd say the most impressive was at Draycote for mid-June to mid-August, 1972. At that time the area had been flooded around three and a half years and the sedge grubs were using the black tubes of rotting straw for their basic home. The bottom simply crawled with aquatic life, and a deep-sunk Black Chenille accounted for scores of brownies in the 2 to 4 lb class. Every one was choked with sedge larvae. Throughout that period an evening hatch was most predictable. It would start around three hours before dark, when winged adults returning to lay their eggs would mix with those newly hatching. Trout interest, desultory at first, would increase as night approached, and would reach a peak about half an hour before dark. In the early stages precise imitation of pupa or adult fly seemed important, but as the fish worked themselves up to a peak of feeding frenzy choice of fly scarcely mattered. They would take practically anything. The sport was superb. Limits were achieved in a matter of minutes, sometimes, before the insects disappeared like magic in the cool of the night. There were two main species of sedge – Silverhorns and Grouse Wing, and a few examples of a smaller and darker sedge which I did not identify. When the sport petered out towards the end of August Draycote did not seem the same, and we were bitterly disappointed when sedge proved sparse in the corresponding period a year later. I do not know why there are good and bad sedge years. It may have something to do with the right combination of conditions, or it may be that the trout decimate the sedge so much in a good year that it takes time to re-establish the numbers. Preoccupation can only occur when the numbers are vast, but most years there are occasional bursts of activity during which we can enjoy one of the most exciting forms of fly fishing.

Daphnia

As you will have gathered earlier, daphnia is a food item of the utmost importance – *the* most important, in the case of Grafham rainbows. It cannot be imitated by the angler, but that does not matter, since we know that trout gorging daphnia are vulnerable to lures. But we need to understand the behaviour of daphnia to gain some idea of where to fish those lures, and with that in mind I took the trouble to read all the scientific works about daphnia. The information I unearthed contributed a great deal to my catches, for the movement of this crustacean member of the plankton family tells us volumes.

The Grafham rainbows practically drink daphnia like soup, and it is beyond doubt that this food is mostly responsible for their remarkable growth rate. Daphnia have a feeding mechanism which enables them to take in water and filter from it minute particles of algae. In the main summer months at Grafham the daphnia blooms are vast, and their reproduction rate in the warm water is astonishing. It probably has to be, to survive the intense predation. In these warmer temperatures algae is also present at all depths, so the food supply is handy.

During countless hours out in boats I have often marvelled at the sight of teeming millions of daphnia in the surface layers, only to scratch my head a few days later and wonder where it had all gone. The answer from the experts is that it continually rises and falls between the surface and the bottom, and the factor which influences that movement is the intensity of light. The natural behaviour of daphnia on a normal day is to ride high in the water early in the morning, sinking slowly into the depths as the sunlight reaches a peak around midday. By mid-afternoon it will be in mid-water, and will rise steadily as the evening wears on. Near nightfall it will be right to the top, and it will stay there through the dark hours until the light after dawn causes the slow descent.

That gives us a rough guide as to where to fish our lures at given periods of the day, but we must also bear in mind that the typical behaviour will change if the day is not typical. A warm but very dull summer day will see the daphnia up top most of the time, and you may remember from an earlier chapter that this is the time when rainbows can be caught on the flashier types of lures – those incorporating the colours orange, yellow and green. Another thing to watch for when the daphnia is high is the influence of the wind. It will push the daphnia to the windward side of a reservoir, and that's when our daphnia soup becomes as solid as Irish stew! Many times at Grafham I have taken advantage of a strong south-wester, which forces daphnia into the funnel-shaped Church Bay. At such times the bay is solid with rainbows, and a limit catch is almost routine.

When the daphnia is not up top, location of the correct depth is not so easy. We have to rely on the rainbows to tip us off. Slow-sinking shooting heads or sinktip lines are the order of the day, and we have to fish them methodically, fishing different depths until a thumping take tells us we have found the right one. Unless you are fishing methodically, however, you will not make the best of it. If you get a take without really knowing how deep the line had sunk before you began the retrieve you will have learned nothing from the experience, and you will have to start again.

The best technique is one adapted from something I read by the late and great Freddie Foster, the swing tip magician who could catch match-winning weights of bream when bites were not registering on his swing tip. If he wound in to find his bait damaged and he had not seen a bite he would assume the bite had come at some stage between the lead hitting the bottom and the bait hitting bottom. To find out exactly when he would count a few seconds and strike, until he made contact with the fish. I can imagine what effect that kind of approach had on the many anglers who watched him! But we can do much the same, though in reverse. We can count down from the top until the feeding depth is located. It may be anything from ten to sixty seconds, or even more, but whatever it is we must find out. The deeper it is the more effective long casting becomes, for the obvious reason that the longer the cast the longer we can keep the lure in the feeding layer during the retrieve.

Perhaps you are beginning to appreciate, now, that the chuck-and-chance-it sneers about lure fishing are quite unjustified. Imitative fishing depends, for success, on the ability to put the right fly in the right place. But there are right and wrong places for lures as well. For the kind of deep fishing just described the important lure colours are black and/or white. Black Chenille and Baby Doll are obvious choices, along with any other combination of white and black, but for the countdown style never underestimate the value of the Missionary and Appetiser. Apart from being the right colour they also go down slowly, like underwater parachutes, particularly when fished on the sinktip. Rainbows have the obliging habit of accepting them on the drop, and there is no better indication of the correct feeding depth, providing you were counting! Find the daphnia and you have found the rainbows is a pretty good motto, though I do not think the message has yet sunk home with the majority of anglers. I was excited by the possibilities when I read what the experts had to say about the behaviour of daphnia, but not fully convinced until I checked it out for myself. I stocked my tiny garden pond with daphnia and watched it in differing light conditions. It behaved as the experts said, and at the brightest period of the day almost the whole lot would retreat into shade provided by a board placed across one end.

Chapter ten

Stocking policies

In the early days of the reservoir trout boom a good many waters made mistakes, but there is now a much better idea of what is required. My main complaint, as I mentioned earlier, is the trend towards stocking with rainbows at the expense of browns. The economic argument is easily understood. Rainbows can be raised to a useful stocking size much more quickly and cheaply and, as much less finicky feeders than browns, they offer better all-round sport. But unless some managements think beyond mere economics, and step up the brown trout ratio, a fascinating aspect of our sport is going to suffer. It would be sad indeed if we lost the challenge offered by big brownies. An 80 per cent–20 per cent bias in favour of rainbows would suit me; two-thirds to one-third would be better still, though.

When waters show the capability to grow good brownies the species should be encouraged. Rutland is my leading brown trout water now, and while Grafham still grows them big it is such an outstandingly good rainbow fishery the managers are clearly correct to stick with a rainbow-orientated stocking programme. Draycote has lost its high position in the brownie league, and is now just another put-and-take reservoir. Stocking levels, as well as a proper brown/rainbow ratio, are a cause for concern in some places. There are times when, to put it as diplomatically as I can, anglers' catches do not appear to correspond at all with what has supposedly been introduced. In many cases which give rise to complaint there are reasons for poor sport which have nothing to do with dishonesty on the part of the managers. Straightforward rip-offs are rare, but not unknown. One department in which some of the Water Authorities show a lack of adventure, however, has been failure to take up some of the newer species. Brook trout have been used to a limited extent, but not the hybrid

Hybrid trout may one day figure more in stocking policies. These cheetah trout, bred at Avington and caught there by Bob Church, weighed between 4 and 5 lbs.

tiger and cheetah trout, which are still confined to a small number of fisheries. This is undoubtedly due to the fairly common policy of Water Authorities to stock waters with their own fish, which tends to deny the small but enterprising fish farmer of the full rewards for his ingenuity. Many farmers regard Water Authority breeding efforts with deep foreboding. Generally speaking they have not yet cracked the problem of mass producing trout at an economic price, but they always have somewhere to put their expensive fish. The national glut of rainbows, which Water Authority production has helped to create, is therefore a problem for the private breeder alone to sort out, and he is entitled to feel aggrieved to see normal commercial reality reversed. It's the fittest who are struggling to survive, though the angler certainly gains little from the situation. The extra cost of Water Authority trout goes on his ticket price on public waters, and they set the standard for the private waters. One result of the glut has been to divert some breeders into producing quality trout for the smaller waters, some of which offer outstanding sport, which they manage to maintain by regular top-up stocking. Places like Avington, Church Hill Farm and Baynham Abbey stock daily, replacing what has been taken out the day before. The big public waters cannot do that, of course, but there has been a big swing to weekly stocking, which has had dramatic results for the angler. Grafham stocks around 3000 fish a week, and the last set of figures I saw, 1980/81, showed that the fishery yielded 85 per cent of its stock, which is tremendous. They would struggle to achieve that, and anglers would have a much more patchy season, if Grafham's 60 000 fish went in all at once or, as happened at one time, in three stages. Staggered stocking maintains a better general standard of sport, and the periodic introduction of 'easy' fish offers great encouragement to the beginner. The unfortunate effect of putting all the year's stock in before the start of the season is amply illustrated by the graph for Chew in 1972 (Fig. 25). It was drawn as part of a more comprehensive study of major reservoirs by a Nottingham friend, Steve Parton. As you can see from the diagram sport slumps alarmingly after a couple of early season peaks, thanks to the stocking policy and, to a large extent, the considerable early season attention of a few anglers and a lot of fishmongers. Since then Chew has switched to a staggered policy,

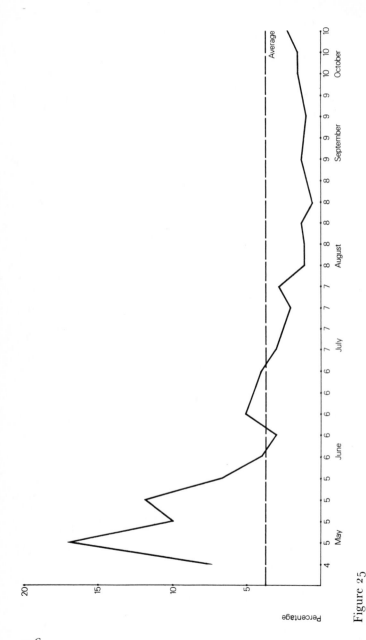

Figure 25

Steve Parton's catch graph for Chew Valley in 1972, showing the percentage of the total catch extracted weekly. Note the nose-dive after May, caused by the policy at that time of introducing the whole of the season's stock before the start of the season.

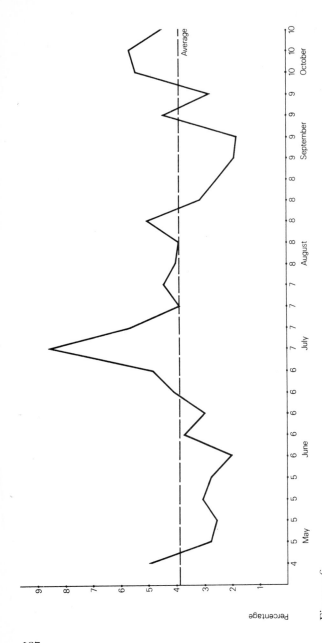

Figure 26

Steve Parton's graph for Draycote in 1972, where the stocking was staggered through the season. The peaks here reflect the stocking times and, to some extent, show how sport improves with the conditions when the stock is kept at a fairly constant level. It would be reasonable to expect a good start in May, followed by below average sport in cold weather conditions. July and August sport in the warmer water is normally good, but anglers usually suffer for their good times for a while until the last restock, and other factors, ginger up sport at the end of the season. All these trends are reflected in the Draycote graph.

Feeding time in the stewponds at Dave Riley's Nythe Lake, Hants.

and I imagine if Steve drew another graph it would probably resemble that for Draycote, 1972 (Fig. 26), which offers a complete contrast.

Studying these and other graphs Steve produced makes me think that if a graph was drawn of an average angler's season it would closely resemble the graph for the water he fished most. It might be an interesting exercise for readers to carry out. If your graph shows catches significantly better than the average you are either a good angler, or well on the way.

Chapter eleven

Favourite fly patterns

The actual mechanics of fly tying is beyond the scope of this book, but if the chance ever arises I would love to devote a separate book to this fascinating subject. Tying one's own patterns, traditional and new, is almost as exciting as catching fish. And unless you are fortunate enough to have an extraordinarily good tackle dealer the flies you can tie yourself will be cheaper and better than any you can buy. Some anglers, of course, are utterly useless at anything which calls for a nimble finger, but the majority are quite capable of putting together a fly or lure which will prove quite acceptable to trout. I reckon I could have the average angler tying a decent lure after a mere twenty minutes' instruction and demonstration. I urge you, therefore, to have a crack at it. Badger somebody who knows how to tie them. Go to night school. Read some of the standard works on the subject. Then you will be able to return to this chapter and tie my recommended patterns in just the same way as those already initiated in the art.

Before I launch into that, though, it is an appropriate moment to discuss the merits of hooks, since tying a good fly on the wrong hook defeats the object of the exercise. I have no connection with any hook manufacturer so the advice is quite unbiased. It is based on long experience – sometimes sad experience. I have had my let-downs with hooks, but I now have the feeling that the quality of some, at any rate, is improving. My favourite hook size for lures is a long shanked 8 or 6, bronze-coloured for most patterns. I use Mustad hollow point No. 9672, which has never let me down, and I also sometimes use the lure hooks brought out by Geoffrey Bucknall in 1975, made by VMC to his design. For lighter-coloured lures I sometimes use nickel-plated hooks, again size 6 and 8. The make this time is Sealey Octopus 1721 N. Sparselydressed flies like nymphs and small wet patterns require a fine but strong hook in sizes 14, 12 and 10, and I always use bronze. I use the popular Partridge type, and can recall only one let-down – when I hooked two rainbows at the same time! For the more heavily dressed wet flies I use a rather sturdier hook – the Mustad 9143 in sizes 12, 10, 8 and 6. An alternative for small wets and dry flies is Geoffrey Bucknall's VMC hook which is forged with a forged eye. The point is excellent and the barb is cut to only 10 per cent of the hook's thickness. The only other hooks I use are small doubles or, as Cyril Inwood and I used to call them, 'wee wets'. They are great hookers when used for small wet flies and they are also useful for hiding in the tail of tandem flies. I rarely use tandems these days, but when I do the rear hook is always connected to the front hook with 20 lb breaking strain nylon.

One example of the Cyril Inwood magic which used to enthral the fly-fishing world. A trout moves near the landing stage at Packington. Inwood casts, and the gallery is treated to a battle with a 5 lb rainbow.

I hear many complaints about the alleged poor hooking quality of various brands, but I will wager the main reason for missing takes is lack of sharpness. Dick Walker has preached the merits of sharpening hooks for years, and the advice goes double for fly fishing. Touching them up with a file before the first cast helps a lot, but I prefer to do a better job at home. I file each side of the hook point when the hook is in the fly-tying vice. Each side is filed flat, transforming the hook into a razor-sharp knife blade instead of a tapering point. It slices into a taking trout and bites deep, even when it hits a bony part of the jaw. It's such a simple precaution; so simple that most anglers forget it. Just because a hook seems sharp when it comes out of the packet does not mean that it cannot be improved upon. If you doubt it try sharpening one and then comparing that hook with an original under a microscope. You will never fish with an unsharpened hook again.

These precautions can often mean the difference between a limit and a brace, though lack of sharpness is not the only factor involved in losing fish. The majority of misses occur when the line is dead straight between the rod tip and the fly. Trout seem to get a better hold of fly or lure when the line is bowed and it is most noticeable that the line which gives the most 'bent' presentation – the sinktip – is the line with which we experience the fewest misses. To recap on another reason for missing, size of fly and speed of retrieve are sometimes at fault. The response to 'tap takes' is to reduce the size of the fly in cold-water conditions and retrieve it a bit faster in high summer.

Now that's out of the way, let's return to the real subject of this chapter – my forty most important flies. I would say thirty-two of them account for almost all my trout throughout a season. Restricted to these patterns alone I would still gamble on taking more than three hundred trout in a season, but I would be struggling to catch a further hundred without the right 'occasional' flies, which are also illustrated. I have already described on the preceding pages how I fish some of these. For those flies which have not already been discussed, I include some remarks on how and when to fish them as well as details of the dressing.

Lures and attractors

BABY DOLL

Use brilliant white nylon wool (Sirdar is a good brand). Take about 9 in of wool and lay it along the hook shank as a large loop facing the eye, leaving an inch-long single strand and a loop of similar length projecting at the tail. Secure both loops and the single strand with tying silk and take the silk down to the eye. Next, take the long single strand of baby wool and loop this over the back of the long loop once. Then build up a long slim body, and secure. Pull the long loop tightly along the back and tie in. Cut off any wool projecting beyond the eye and build up the head with black tying silk. Cut the tail loop to about half an inch and then, using a needle, shred the three strands and cut into a bushy tail.

This fly was created by Brian Kench in 1971.

WHISKY FLY

Invented by Albert Whillock, this pattern is another of the real success stories in modern reservoir fly dressing. It is tied in a number of variations, the one I favour being as follows. The tail is orange hackle fibres and the body silver tinsel. Orange hackle fibres form the throat hackle, and four orange cock hackles form the wing.

MUDDLER MINNOW

The creation of Don Gapen of Minnesota, who invented it for fishing the Nipigon River in Northern Ontario.

The tail is oak turkey feather, and the body is flat gold tinsel. The wing is grey squirrel hair and oak turkey feather fibres. The shoulders are natural deer body hair spun to surround the hook and then clipped to the required shape. A few of the deer hair fibres sloping back from the rear of the head should be left unclipped to give a hackle effect.

JACK FROST

The tail is a small segment of red fluorescent wool. Underbody is white wool, as used in Baby Doll, covered with stretched polythene. The wing is a generous spray of white marabou herl. For the hackle use two full turns of red and three full turns of white. Make sure these hackles have long fibres, and tie them down so that they sweep back towards the tail.

One of my own creations, in 1974, this lure is particularly effective when trout are on the fry. The marabou wing helps to give it life, and it can be fished very slowly to good effect. It seems to work on all waters where trout feed on fish fry.

BLACK CHENILLE

Tail is black cock hackle fibres and body is black chenille wool, spiralled with silver tinsel. Throat hackle is black cock hackle, the same as the tail, and four black cock hackle feathers form the wings.

One of my own inventions, first tied in 1970.

APPETISER

Tail is slightly thicker than usual, and made from mixed orange and dark green hackle fibres, and just a few fibres of silver mallard feather. The body is white chenille ribbed with silver tinsel, and the throat hackle is made up from the same materials which form the tail. The wing is formed in two layers – first a generous spray of white marabou feather followed by a spray of natural grey squirrel tail. You might also use black squirrel tail or black goat hair as an alternative.

Another of my own patterns, first tied in 1973.

CHURCH FRY

Tail is white feather or hackle fibre and the body is orange chenille spiralled with gold tinsel. Orange fibres form the throat hackle (red is a good alternative) and the wing is grey squirrel tail (natural).

I introduced this pattern in 1963.

SWEENEY TODD

No tail. Body is black floss spiralled with fine silver tinsel. The throat is three turns of magenta fluorescent wool, and the throat hackle is a spray of magenta cock fibres. Black squirrel tail hair forms the wing.

Richard Walker developed this dressing a few years ago, his idea being to transfer the known attractor qualities of a red tail to the throat region in an attempt to overcome short takes. An effective killer on all waters, and especially in the early season.

BADGER MATUKA

Body is white chenille, ribbed with silver, with a short segment of red fluorescent wool immediately behind the head. Do not rib the body until you are ready to tie in the wing, which would be better described as a kind of dorsal fin running from head to tail. This fin is achieved from two or three pairs of well marked badger hen hackles, which should be matched – i.e. dull side to dull side. The fibres from the bottom side of the feather should be torn off, apart from a short piece at the thin end, which forms the tail. Tie the feathers in at the head and, keeping the feathers perfectly upright with the fingers, fix them firmly to the top of the body with five well-spaced turns of the silver tinsel. Take it through the fibres down to the stalks of the feathers, trapping them very firmly to the body. The spirals of tinsel go the opposite way to the normal tying. The throat hackle is orange cock hackle fibre.

Introduced by Steve Stephens in 1972, this lure swims straight as an arrow and begs for a wetting when the trout are taking fry. More of a fry imitator than a mere attractor, it often works when fished deep for brownies.

MISSIONARY

Tail is a spray of red cock hackle fibres and the body is white chenille spiralled with silver tinsel. The throat hackle is the same as the tail, and the wing is silver mallard feather kept whole, and tied so that the quill lies parallel with the hook shank. An alternative wing is barred teal feather. Another alternative, which gives the lure a more streamlined appearance, is to tie the wing with a bunch of barred feather torn from the quill. The Shrive pattern is amazingly good for catching rainbows on the drop. The alternative works better with a faster retrieve. A final variation for both patterns is to have white cock hackle fibres for tail and throat.

An old Blagdon pattern which, although famous in New Zealand, had been virtually forgotten in Britain, until Dick Shrive's modern version in 1960.

ACE OF SPADES

Body is black chenille and the wing is black cock hackle tied in with oval silver tinsel in the manner described for the Badger Matuka. The oval wing is bronze mallard feather fibre and the throat hackle is guinea fowl fibre.

A variation of the New Zealand Matuka patterns, designed by Dave Collyer, this is a fine alternative to such lures as the Black Chenille when the fish show a preference for black. Its main asset is that it swims beautifully, thanks to the Matuka-style wing. Deceives well when fished slow.

POLY-BUTCHER

Tail is red cock hackle fibres. Underbody is either red tinsel or red silk (one layer of whichever it is) overlaid with clear polythene, which must be stretched as you work to achieve a smooth and solid finish. The throat hackle is red cock hackle and the wing is black squirrel hair. Since I first made this lure I have added two refinements – a spray of black marabou to the wings, finishing off with jungle cock eyes.

I was thinking of a modern version of the traditional Bloody Butcher when I designed this lure in 1972. The polythene body makes it extremely durable, and I once caught thirteen trout on one and put it back into the wallet unmarked. It seems to have the curious capacity to catch when other lures won't. I try it when I'm baffled, and it often breaks my duck.

MICKEY FINN

Silver tinsel body. Wing is yellow goat hair topped by four slender red cock hackles. Throat hackle is red hackle fibre.

An American pattern, specially effective over there, as here, for rainbows.

MRS PALMER

Tail is white hackle feather fibres and the body is white chenille spiralled with silver tinsel. White fibres form the throat hackle and the wing is yellow goat hair.

Invented by Richard Walker in 1973, this lure is an out-and-out attractor which works well when fished fast in the main summer months. It is also one of the best dirty-water lures, since fish seem to see yellow in murky conditions better than almost any other colour.

POLYSTICKLE

The body is a strip of stretched polythene wound up and down the hook, with a few turns of red floss silk at the throat. The back and tail are formed with a strip of brown raffene which is moistened and then tied in with black silk at the tail. The raffene is then stretched towards the head and tied in there. The tail is spread to a fan shape and trimmed with scissors. Throat hackle is orange fibre. When varnishing the head also treat the tail and back sections. This lure is best tied on a nickel-plated hook.

This lure was introduced by Richard Walker in 1966, and you either love it or hate it. It seems to work for some and not for others. Cyril Inwood, believe it or not, never caught a single fish on it, but my experience is the reverse. It slaughtered trout for me in the early days at Grafham, when sticklebacks were the only small fish for trout to feed on, and it works for me today on stickleback waters. I fish it high and jerkily on a floating line, attempting to make it look like a wounded stickleback.

LEPRECHAUN

Tail is lime-green hackle fibre and the body is fluorescent green chenille, ribbed with silver tinsel. Throat hackle is lime-green

PLATE 4
Muddler Minnow
Church Fry
Baby Doll
Polystickle

PLATE 5
Invicta   Mallard & Claret
Olive Quill   Ginger Quill
Silver Butcher
Grey Duster   Tups Indispensable
Pond Olive

PLATE 6

Brown Buzzer	Black Buzzer
Green Buzzer	
Claret Hatching Midge	Green & Brown Nymph
Black & Peacock Spider	Pheasant Tail Nymph
Floating Snail	Flying Ant

PLATE 7

Spent Mayfly

Stick Fly

Daddy Longlegs

Brown & Yellow Nymph

NOTE: The Soldier Palmer, Thicket, Brown Murrough and Renegade (Plate 8) are not mentioned in Bob Church's top 40 list. They are all loch flies (see loch fishing chapter).

PLATE 9 A dry bi-visible Sedge.
How a dry fly should sit on the surface if it is properly tied and treated with a good floatant. This fly was treated with a readily available liquid floatant which makes it float permanently after just one immersion.

PLATE 8

Soldier Palmer   Thicket
Brown Murrough
Standard Sedge   Renegade
Drone Fly

hackle, like the tail, and the wing is four lime-green cock hackles.

Lime-green, like orange, seems to have special attractor qualities when trout are in the mood to chase something fast and flashy. Concocted by Peter Wood in 1972, the Leprechaun is particularly successful when there is a lot of algae up, and it is another lure with the capacity to catch when other lures are refused.

WHITE MARABOU MUDDLER (Tandem)

Rear hook ties as for Baby Doll, but add a wing – a generous spray of white marabou feather. Front hook is again the Baby Doll tying, leaving enough space near the eye for a Muddler head of white deer hair. The wing is white swan topped with white marabou. Then spin on the deer hair head and clip, leaving plenty of bulk. The overall length of this lure is 3 in but you can go up to 4 in.

I invented this outsize lure in 1973 for use when spooning a trout indicates that they are taking large-size fry.

Note
This completes my list of key lures, all of which are tied on No. 6 or No. 8 long-shanked hooks.

Traditional small wet flies

CLARET & MALLARD

Tail is tippet and the body claret seal's fur ribbed with fine gold tinsel. Hackle is two or three turns of claret cock, and the wing is bronze mallard feathers.

One of the best-known traditional loch type of wet flies, said to have been invented by William Murdock of Aberdeen.

OLIVE QUILL

Tail is four fibres of medium olive cock hackle and the body is stripped peacock quill dyed olive. The hackle is two or three turns dyed medium olive, and the wing is dark starling wing feather.

A wet-fly adaption of F. M. Halford's famous dry fly.

BLACK & PEACOCK SPIDER

Body is four or five strands of bronze peacock herl reinforced with a spiral of fine fuse wire or fine tinsel. Hackle is full circular long-fibred black hen hackle, tied to slope backwards.

One of Tom Ivens's patterns.

GINGER QUILL

Tail is four fibres of pale brown-ginger hackle, and the body stripped peacock quill dyed brown-ginger. Hackle is full circular, pale brown-ginger, and the wing is pale starling.

Another wet-fly adaption from Halford's series of quill-bodied patterns.

BUTCHER

Tail is red ibis and the body flat silver tinsel, ribbed with thin oval tinsel. Hackle is black cock, and the wing is blue-black feather from a mallard drake's wing.

One of the best-known traditional wet flies, said to have been invented by a Mr Jewhurst of Tunbridge Wells in the early nineteenth century.

Note

All these small wets are tied on medium-shanked hooks in sizes 14, 12 and 10.

CHIRONOMID PUPAE (buzzer nymphs)

The three very important body colours are black, brown and green. All three are tied exactly the same, apart from the colour of the wool. Start the body half-way round the bend of the hook, using only one strand of wool and keeping it very slim. Spiral with a very narrow strip of tinsel or fuse wire. Form the thorax with one or two strands of bronze peacock herl, according to hook size. The cases are my refinement of the standard dressings – two starling hackle tips which add the perfect finishing touch.

CLARET HATCHING MIDGE

Body is claret wool spiralled with fine silver tinsel. Thorax is peacock herl, and the hackle two full turns of honey-coloured cock hackle.

This is my version of the oldest of the buzzer patterns, still very effective when fished just under the surface during a buzzer hatch. It never seems to work fished deep, but up top it looks very much like a midge pupa on the point of emerging.

PHEASANT TAIL NYMPH

Tail, pheasant tail feather fibres. Body is pheasant tail fibres wound round the hook for two-thirds of the length back from the hook towards the eye. For the thorax tie in some more pheasant tail fibres pointing towards the bend of the hook. Now tie in one strand of peacock herl and build up a ball shape. Overlay this with the loose fibres which were pointing towards the hook bend, taking them tightly over towards the eye, and tie off.

This is one of the top fish catchers among the imitative patterns, though it is intended to imitate a variety of nymphs rather than one. It is the favourite of one of the top Midlands nymph fishers, Arthur Cove, who fishes it deep and slow with floating line and long leader.

Note

Hook sizes for the chironomid pupae and the other two nymphs should be 14, 12, 10 and 8.

Sedge Nymph Patterns

BROWN & YELLOW NYMPH

Tail is pheasant tail feather fibre and the first two-thirds of the body, starting from the hook bend, is brown dubbed seal's fur. The remaining third is yellow seal fur, and the whole body is ribbed with narrow gold tinsel. The hackle is two full circular turns of short-fibred yellow cock hackle.

There are a number of similar patterns, the common factor being the combination of body colours. This one was passed on to me by John Wilshaw.

GREEN & BROWN NYMPH

Body, green wool or dubbed seal's fur for two-thirds of the hook length, the rest bronze peacock herl. Hackle is two full circular turns of brown cock hackle.

The invention of Vic Wright.

STICK FLY

Body is four to six strands of peacock herl spiralled with fuse wire, with one turn of pale yellow wool near the head. Hackle is four turns of honey-coloured cock.

This is my version of the standard pattern.

STANDARD SEDGE

Tail is pheasant feather fibre and the body is fibre from a dark cock pheasant tail wound on thickly. The hackle is full circular ginger cock and the wing is natural deer hair tied in by the tips and trimmed level with the hook bend. Leave splayed out.

My version of one of Terry Thomas's sedge patterns.

INVICTA

Tail is golden pheasant crest-feather fibre and the body dubbed yellow seal's fur ribbed with oval gold tinsel. Body hackle is red game from shoulder to tail and the front hackle is blue jay wing. Wing is from hen pheasant's centre tail.

Invented early this century by James Ogden of Cheltenham.

Note

All the sedge family patterns are on Nos. 12, 10 and 8 hooks, but the Brown & Yellow Nymph and Stick Fly are tied on No. 8 long shanks.

Miscellaneous dry flies

TUP'S INDISPENSABLE

Tail is honey dun hackle fibre and the lower half of the body is yellow floss. Above that are two turns of mixed red and pink seal's fur dubbing. Hackle is two full turns of badger cock.

A much-modified version of the famous dry fly invented by R. H. Austin at the turn of the century.

GREY DUSTER

Tail is optional, but if used should be of honey-coloured fibre. Body is grey rabbit fur with a little blue seal fur dubbed on. Hackle is well-marked badger cock, with dark centre and light tips.

An adaption of a Welsh river pattern.

Note

These two flies are tied on hooks Nos. 16 and 18. I mostly use them when trout are rising to caenis. If they are refused dry, then allowing them to sink just into the surface film and fishing them very slowly sometimes works.

SPENT MAYFLY

Tail is pheasant tail fibre and the body is white foam plastic with two black bands made with the black tying silk as you progress along the hook. The wings are four dyed violet-blue cock hackles, two on each side and tied spent fashion. Hackle is four turns of grizzle, long-fibred, followed by two turns of shorter badger fibres.

My own variation – 'unsinkable' by virtue of its buoyant foam body. Hook size: long-shanked 8.

POND OLIVE

Tail is a few fibres of olive cock hackle. Body is two strands of pheasant tail fibre pulled tight. Hackle is selected olive cock, tied circular and the wings are light starling, tied upright.

Hooks: 14, 12 and 10.

DADDY LONGLEGS

Tail, or in this case the legs, is four pheasant tail fibres tied to protrude one inch. Body is a thin strip of cork reinforced with fuse wire. Into the shoulder tie two more long legs of pheasant tail fibre (two from each side). They should be about two inches long and knotted about half an inch from the tips. The hackle is ginger cock and the wings, spread straight out, are four ginger cock hackles, two each side.

One of the late Cyril Inwood's patterns. Hook: No. 10 long shank.

DRONE FLY

Use red tying silk. Body is yellow chenille spiralled with one strand of bronze peacock herl. Wings are two white cock hackle tips and the hackle is full circular, medium brown. Tie off with a red head.

Another pattern invented by Cyril Inwood. Hooks: 14 and 12.

FLYING ANT

Rear end of body built up egg-shaped with black silk or wool. Go into a narrow waist midway along the hook and then create a slight bulge again near the bend. The wings are two starling hackles and the hackle is black cock.

One of Dave Collyer's dressings. Hook: 12

FLOATING SNAIL

The easiest of all to tie. It is simply black or brown chenille tied round the shank in a big ball shape. I tie some of each colour on 12 and 10 hooks for fishing just beneath the surface film on the occasions when trout are taking migrating snails.

Random thoughts on flies and tying

Several things occur to me as I gaze at the box of 40 flies, all of which were specially tied for the illustrations in this book. The collection looks meagre. It occupies a mere few inches of plastic foam – 17 lures, of which five (Whisky Fly, Church Fry, Mickey Finn, Mrs Palmer and Leprechaun) can be classified as Attractors and the rest Deceivers; 8 dry flies, all Deceivers; and 15 wets, again all Deceivers apart from the Butcher, which is an Attractor. Attractors, of course, are flies or lures which are intended to resemble nothing which occurs naturally in or on the water. They are without exception gaudy, flashy creations designed to trigger the aggressive streak in trout. They work mostly when the trout are feeding freely, and sometimes they will provoke a take on dour days when not much is happening. The Deceivers are either close copies of aquatic or land-born creatures or rough approximations of fish or fry. I can visualise some of the fly-tying experts shaking their heads sadly as they pick over my collection, and mentally ticking off the 'vital' omissions. I can also visualise other writers in the more recent past encountering the same feeling, and giving way to the temptation to expand into the more erudite backwaters of the fly fishing art. Almost without exception such expansion has complicated the picture to a degree which baffles the novice and adds little or nothing to the knowledge of the expert. I have no intention of falling into the same trap. Inventing a few successful flies is the limit of my vanity.

Fly fishing can be likened to stamp collecting. Some people buy stamps and keep them in albums while others buy them and stick them on envelopes. The fly fishing parallel is to tie flies for the admiration of other fly tiers or to tie them to catch fish. If you set out to tie every fly that ever caught a trout you would never go fishing. The game is so fascinating you have to watch that it does not become the primary pastime; that would be as silly as abandoning tennis in favour of stringing rackets. Fly fishing and fly tying is so rich in literature that any tyro who takes the precaution of reading the classics is liable to confuse himself so much that he will switch to something more simple, like chess. Angling must be the only game in which reading the standard works is more educational and entertaining *after* one has reached a degree of competence through one's own efforts and through close association with expert anglers. That way one becomes able to understand and appreciate what the great writers were saying.

To be successful these days the first requirement is to master the various methods of fishing and to come up with a compromise collection of flies and lures. The latter is no mean task, given the impossibility of being comprehensive. Out of a veritable jungle of fur and feather we have to hack a handful which will cope with requirements 90 per cent of the time. My meagre collection amounts to my compromise, though meagre is a deceptive expression. I think I can claim that my record over the years amply establishes that my compromise is effective, and that it meets most requirements. When you consider the multiplication of sizes which are involved, and the number of examples required on each size of hook, my collection no longer looks meagre. Half a dozen of each size gives you 240 flies. And if we assume that, on average, there are three different sizes for each pattern you have a wallet full with 720! Every time you add a pattern the numbers rocket, and clearly there is a point at which you have to cry enough. I am quite happy that my range of lures is right, and that the sedges, buzzers and small wets will see me through almost all the periods in which the trout are preoccupied with natural insects. The sprinkling of occasional flies fills some of the remaining gap, and there I rest my case.

I accept that I can never be armed for every emergency which may arise; nobody can be. My only reservation is that my effective list of flies and lures may not be quite so effective for you as it is for me, for although my fly fishing parish is extremely far flung I do not fish everywhere. Most of them will catch fish anywhere, but there is always the possibility that a particular water will hatch off something the trout like which I have not mentioned. It is up to you to get to know your own water and arm yourself with the right imitation.

You may find your local sedges are a different colour, for there are regional variations in colour, darker or lighter. In addition the sedge family is vast, and different types may occur more profusely in some areas than in others. I well remember a sedge rise at Grafham during which my imitation proved about half as acceptable as the deer-hair version being fished by Terry Thomas, which is illustrated in this book. I have remained converted to his tying ever since, but my old pattern is still tucked away in the wallet somewhere in case I find myself among strange sedges where Terry's won't work.

Similarly the buzzers belong to a huge family, and whilst I have always found black, brown and green to be the effective colours there are great regional variations in size. Chew, for example, hatches some enormous buzzers, so I always carry a small collection of outsize examples for possible use at Chew alone.

An angler who gets to know his waters well soon winds up with a range of flies which will serve him for most eventualities, and he also arms himself for the occasional but predictable periods of trout preoccupation. As the years roll by that corner of his wallet which houses the occasional flies tends to get a bit crowded, but the collection of old faithfuls remains much the same. There may be the odd improvement in the tying of a particular fly and, now and again, another 'must' for the lure collection, such as the Baby Doll or the Appetiser. There may also be change in the materials or, to be more precise, a new material which does a job better than a traditional material or which adds to the appeal of the old. Into that category, certainly, falls the marabou feather and the new fluorescent materials. I use marabou on four of my key lures, and I am the first to admit that it does nothing for the appearance of the lure when it is fixed in the wallet. My Poly-Butcher and the White Marabou Muddler in particular resemble ridiculous balls of fluff. But the penguin looks ridiculous on dry land, yet in the water. . . . Watch a marabou lure in water. It literally pulsates with life as the fine feather reacts to the movement the angler imparts. If anyone tells me that this does not help to catch fish then they are also promoting the heresy that exact imitation has no place in fly fishing. But how exact is exact? We can get pretty near it with a fair number of flies, but the greatest fly tyer who ever lived would baulk at the suggestion that he had achieved perfection. Only mother nature achieves perfection, and she is a hard lady to copy. Examine any insect and you will readily understand the sheer impossibility of exactitude. It is not just the delicate tracery of the wings and the fragile grace of the legs which defeat you; it is colour as well.

A fly which appears to be one colour is really a kaleidoscope of colours which blend together and suggest there is just one. We were stuck with using that one colour until the introduction of DFM, which does not stand for Distinguished Fishing Medal! It means daylight fluorescent material, which has quietly re-volutionised fly tying. We can now have lures which are whiter than white. The fluorescent white wool which forms the Baby Doll is a key reason why this lure kills an undue share of trout. Its bright translucence is so important I discard my Baby Dolls after one day's use, for they collect dirt and lose that inner glow. When we come to the colours DFM has enlivened some of the old traditional fly patterns, and it has enabled us to produce such gaudy rainbow killers as the Church Fry and the Leprechaun. For the occasions when the trout won't take the flies which look like aquatic fireballs we can tie lures which use DFM more moderately, but which still transmit that fatal flash on a fastish retrieve. Toning down the use of DFM still further we can achieve a mere suggestion of difference in that one colour; we can get some way towards mother nature's kaleidoscope of colour. There is immense scope for experiment, still, with DFM. But until somebody proves otherwise I will stick with a simple

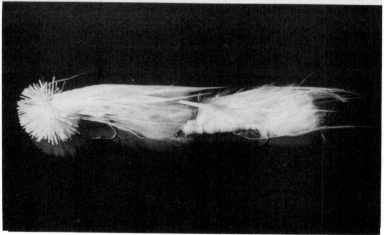

The White Marabou Muddler in dry form may appear to be a monstrosity (*top*), but the bottom picture gives an idea of how it slims down to a more fishy shape when wet. When actually fished, the marabou feather gives the lure a pulsing, lifelike quality which helps to deceive fry-feeding trout.

formula of my own. For lures which are fished fast I either use a lot of DFM to create a real flasher, or one part DFM to three parts traditional material. For the much more subdued effect I reckon one part DFM to five traditional is about right. When we come to the colours which are most effective I have concluded that lime green, orange, red, white and magenta are best, with yellow the colour best seen in dirty water.

Even since the introduction of DFM, fly tying materials have continued to improve. One major advance has been forged by necessity. Seal's fur has been banned in America, giving rise to outstandingly good synthetic dubbing materials for nymph and wet fly bodies. There is a greater variety of colour, and there will be no need to ban seal's fur in Britain; it will be gradually phased out by the swing to synthetics. Chenille has improved too. There's now a sparkle chenille which has lurex tinsel blended in, and a very fine suede chenille which we now use for some nymphs and wet flies.

Judging from the stack of mail which inevitably results from regular writing for various angling publications my lures catch plenty of fish for other people. The only criticism I get from time to time has nothing to do with the tying or the design. It revolves around the number of lures and the relatively small ratio of deceivers. My answer is that the collection reflects reality. Most of the time the lure is the likeliest method of catching fish.

If there is no visual indication leading to the correct choice of a deceiver what is one supposed to do? Guess what the fish might be taking and fish more in hope than expectation? Obviously not. The answer is to fish an attractor with known fish-killing capability on the water concerned. If a pal does the same with another attractor of contrasting colour, one or the other, or sometimes both, will catch eventually, and spooning the fish should then produce a sound indication of what to do next. It might be to carry on with a lure or to switch to the appropriate deceiver. When there are surface signs the question of choosing the correct deceiver is rather more simple. I would have thought that this general approach is the sensible one for reservoir trouting, but it does lead one into what some seem to consider excessive use of the lure. I do not understand this line of thought.

I use more lures than deceivers simply because knowledge, experience and observation tell me that lures are required more often than deceivers.

I hope I have proved in the passages that have gone before that lure fishing is a highly skilled business and not a haphazard, chuck-and-chance-it exercise. But that is how it is condemned by some, who mouth the invective 'lure stripper' in tones usually reserved for 'child molester'. The big swing to reservoir fishing helped to combat snobbery, but it is not yet dead. It lives on mostly among the chalk stream converts, some of whom appear to think that to catch a trout on anything except nymph or dry should be a capital offence. They are, of course, entitled to their opinion, but it has no basis in common sense. It is unrealistic to take chalk stream methods and chalk stream ideology and try to superimpose them on the reservoir game. The two branches of the sport are totally different. Beginners seem to evolve a sensible approach to reservoir fishing faster than the converted river men. Maybe it is a reaction to the discovery that their many years of experience on rivers amounts to nothing more than mastery of a tiny fraction of the overall technique required for success on reservoirs. They then fall into the old trap of condemning that which they do not understand and, having condemned it loudly and long enough, they are prevented from becoming successful all-rounders by stubborn pride.

It is not simply lure fishing which wrinkles a few aristocratic noses, either. Boat fishing is much misunderstood, and the author of one recent book actually wrote it off as little more than an easy way for novices to catch fish! I think this book has already demolished that idea. Let a novice loose with a boat and he is likely to catch far fewer fish than he will catch from the bank. Reservoir anglers have to make up their minds what they want to do – dance in the chorus or play the lead. I buy my ticket and that great big watery world is mine. All of it. I have bought it from sunrise to sunset, and I want all it has to offer, not just part of it. I do not think how I would *like* to catch fish, and take it from there. I just set about discovering how it is *possible* to catch fish that day, and get on with it. The last thing I would do is sit down and weep because the sedge didn't hatch, the olives took a holiday and the buzzers buzzed off. Neither would I dump a lure-caught limit on the scales on such a day and feel anything but the ultimate satisfaction. I certainly would not look at the biggest brownie, which might have chomped a Marabou Muddler in 30 feet of water, and wish that I had caught it from concealment in a reed bed as it made an early morning stickleback raid. And the big rainbow which grabbed a Mickey Finn is not diminished because it did not suck in a neatly cast Olive. I take these and all other delights of reservoir fishing as and when they come. Nobody can manufacture magic moments. All we can do is enjoy them when they happen, and for me the magic is encapsulated in that moment when the trout takes my fly – whatever that fly is and wherever it is fished. I pity those who see shades of magic, white and black, and limit themselves to the white. I reserve another emotion altogether, though, for those who seek to make a virtue of their limitations and impose their prejudice on others.

The rules I live by are printed on the fishery lodge wall and on the back of the ticket. I care nothing for unwritten laws, especially those laid down for another world altogether. Study the reservoir returns, and you will see well enough what we are up against. The average catch is around one trout per rod per day. My average last season, on eleven reservoirs and four smaller lakes, was seven fish per rod per day. I am seven times more successful than those who dance in the chorus, and if that sounds brash and boastful then I'm sorry, because it is not intended. Some of my friends do even better than me at times, but you don't catch me muttering behind my hand about their methods. I buttonhole them and ask what they are doing, and more often than not a new piece fits into that never to be completed jigsaw. The reservoir world is an open air university, a never-ending degree course for which the diploma is always at the end of the rainbow. So much to learn . . . so little time. . . . How can anybody, in those circumstances, treat reservoir fishing as a kind of gentleman's handicap race?

Let me put the situation to you as bluntly as I can. If you are averaging a fish per rod per day you are paying for my fish, just as others like you paid for the vast tonnage which fell to Cyril Inwood. The challenge which faces you is to buckle to and get someone to pay for yours. To attain that status is within the compass of any man who tries hard enough, and he who tries can get all the help he needs. Weigh yourself down with fish, and not the albatross of prejudice.

Index

Note: numbers in italics refer to illustrations and figures

Index